Fred H. Whipple

Municipal Lighting

Fred H. Whipple

Municipal Lighting

ISBN/EAN: 9783337249625

Printed in Europe, USA, Canada, Australia, Japan

Cover: Foto ©berggeist007 / pixelio.de

More available books at **www.hansebooks.com**

The Electrical Supply Company,

171 Randolph Street, CHICAGO.

MANUFACTURERS OF

Bare-Insulated Wires-Cables

Of Every Description.

TOOLS, ALL KINDS,

TEST INSTRUMENTS,

CONSTRUCTION MATERIAL.

EVERYTHING REQUIRED FOR

The Construction or Operation of Electric Light Plants.

Write for Catalogues.

Factories:
ANSONIA, CONN.

Branch Offices:
125 W. SIXTH ST., KANSAS CITY.

The Best Lamp!   *The Best Dynamo!*

THE UNITED STATES ELECTRIC LIGHTING CO.

— MANUFACTURE —

Incandescent and Arc Light Apparatus

FOR ISOLATED PLANTS

OR CENTRAL STATION USE

AND

CITY OR MUNICIPAL

LIGHTING

CLAIMING

SUPERIORITY OVER OTHER SYSTEMS

BECAUSE OF

Economy of Power.
Precision and Reliability of Measuring Instruments.
Durability of Lamps.
Absence of Discoloration of Lamp Globes.
Perfection and Automatic Regulation
Mechanical Design and Workmanship
Completion of Details.

C. C. WARREN,

MANAGER WESTERN DEPARTMENT,

THE ROOKERY, 219 LA SALLE ST., CHICAGO.

General Offices, Equitable Building, 120 Broadway, New York.

FOR STATION LIGHTING purposes we are furnishing our new and improved Alternating Current System by which lights can be operated at long distances from station, with small cost for conductors.

SEND FOR PRINTED MATTER.

CENTRAL ELECTRIC CO.

Electric Supplies of every description in stock
at bottom prices.

WESTERN AGENTS  OKONITE Wires and Cables.

CANDEE WEATHERPROOF

Electric Light Line Wire,

National and Sunlight Carbons

FOR ANY SYSTEM.

Butler Hard Rubber Co's productions of

Tubins Rod and Sheet Rubber,

CLEVELAND INCANDESCENT

Gang Switches & Arc Light Cut Outs,

CONSTRUCTION TOOLS,

Shovels, Spoons, Digging Bores, Augers, Line
Wire Reels, Soldering Pots, etc., etc.

———MANUFACTURERS OF———

PINS AND BRACKETS.

HEADQUARTERS FOR

Insulators, Cleats, Tape, Compound, etc., etc.

Orders carefully and promptly executed.

CENTRAL ELECTRIC CO.,

42 LA SALLE ST.,

CHICAGO.

PAYNE HIGH SPEED
CORLISS ENGINE
SHAFT GOVERNOR COMBINED WITH CORLISS WRIST-PLATE.
Economy of Fuel.

☞ REGULATION EQUAL TO ANYTHING IN USE.

B. W. PAYNE & SONS,
ELMIRA, N. Y.

10 S. Canal Street, CHICAGO, ILL. | 45 Dey Street, - - NEW YORK.

HILL, CLARK & CO., Boston, Mass.

AUTOMATIC ENGINES FROM 2 to 200 HORSE POWER.

Municipal
Electric
Lighting
is Feasible
and
Economical
by
the Use of

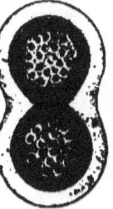 

THE [Electric Light Cable, 2 Conductors. No. 1, B. & S. G. Full size.]

WARING ELECTRIC LIGHT CABLES
MADE BY THE
Standard Underground Cable Company.

General Offices, 708 Penn. Ave., PITTSBURGH, Pa.
Branch Offices—16, 18, 20 Cortland St., New York;
G. L. Wiley, Man'gr; 139 E. Madison St., Chicago, Ills., F. E. Degenhardt, Man'gr.
Manufacturers of the Waring Anti-Induction and Bunched Cables, for Telegraph, Telephone, Electric Light and Power; Underground, Submarine and Aerial.

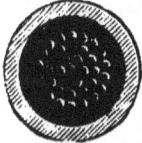

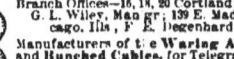
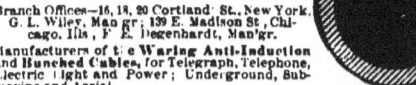
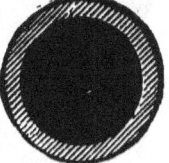

Weatherproof Line Wire, W. A. C. Fire & Waterproof Wire, Annunciator and OFFICE WIRE.

FIVE
YEARS OF
UNIFORM
SUCCESS.
Correspondence
Solicited.

Full size of No. 1. Standard Weathe-proof Line Wire.

THE
Van Depoele System of Electric Lighting

IS SPECIALLY ADAPTED FOR

MUNICIPAL LIGHTING.

THE ONLY PERFECT DOUBLE CARBON LAMP IN THE WORLD.

VAN DEPOELE APPARATUS

Requires the Least Attention.
Is Self-Regulating and Simple.
No Cumbersome and Unsightly Attachments.
No Danger from Handling.
Largest Light for Smallest Power.

The Van Depoele System has by actual test shown that its lamps light the largest area of any lamp in existence, and with the smallest proportionate consumption of power. It is specially adapted for Street Lighting. Our references are all who have ever seen them.

3,500 LAMPS NOW IN USE.

☞ For Catalogues and Information address

VAN DEPOELE ELECTRIC MANUFACTURING COMPANY,
15, 17, 19 & 21 No. Clinton St., CHICAGO, ILL.

RUSSELL & CO.

MASSILLON, OHIO,
BUILDERS OF

Automatic Engines
BOILERS, ETC.
Complete Power Plants Furnished and Erected.

SEND FOR CATALOGUE.

The Clark Electric Co.,

NEW YORK

The Wellington Belt Holder

Just the thing for shifting Dynamo Belts. Holds belt at rest and relieves tension. Simple and Cheap.

                LYNN, MASS., March 2, '85.
W. R. SANTLEY & CO.:
    Gentlemen—After subjecting one of the Wellington Belt Holders to the test of over a year's practical use in our testing room, we were so well satisfied that we have recently purchased five more for the same purpose. We can cheerfully recommend the holder as being a reliable, efficient and at the same time inexpensive article.
    Yours respectfully,
           THOMSON-HOUSTON ELEC. CO.

The above are sending us repeated orders, and recently ordered a Holder for shifting a belt from a ninety-six inch by twelve driving pulley.

Circulars and further information will be furnished by

W. R. SANTLEY & CO.,
WELLINGTON, OHIO.

250 HORSE POWER HEINE SAFETY BOILER.

HEINE ✢ SAFETY ✢ BOILERS

Are in Use in the following Electric Light Plants:

Allegheny County Light Co.,	Pittsburgh, Pa.
Boston Edison Station No. 2,	Boston, Mass.
Chicago Edison Co.,	Chicago, Ills.
McVicker's Theatre,	" "
Interstate Exposition,	" "
Minneapolis Industrial Exposition,	Minneapolis, Minn.
Forest City Electric Light and Power Co.,	Rockford, Ill.
Colorado Electric Co.,	Denver, Col.
St. Louis Exposition,	St. Louis, Mo.
Brush Electric Light & Power Co.,	Galveston, Tex.
Little Rock Electric Light Co.,	Little Rock, Ark.
Columbus Electric Light & Power Co.,	Columbus, O.
Roe Building,	St. Louis, Mo.

For Full Description and Prices write to

HEINE SAFETY BOILER CO.

102 North Main Street, ST. LOUIS, Mo.
82 Madison Street, CHICAGO, Ills.
55 Oliver Street, BOSTON, Mass.

THE THOMSON-HOUSTON

Electric Company,

MANUFACTURERS OF

ELECTRIC LIGHT MACHINERY

Arc and Incandescent

DYNAMOS AND MOTORS.

CONSTRUCTORS OF

Electric Railways and Tramways.

PRINCIPAL OFFICES:

178 Devonshire St., BOSTON, MASS.

Pullman Building, CHICAGO, ILL.

115 Broadway, NEW YORK.

Kimball House Build'g, ATLANTA, GA.

HAMILTON-CORLISS ENGINE

Made in all sizes, from 30 to 1,000 Horse Power.

NON-CONDENSING,
CONDENSING,
COMPOUND,
HORIZONTAL or VERTICAL,
SINGLE or IN PAIRS.

We guarantee our Engine equal to any made in Economy, Strength, Weight or Finish, and solicit correspondence.

THE HOOVEN, OWENS & RENTSCHLER CO., Hamilton, Ohio.

The Waterhouse System

GOLD   **POWER.**   MEDAL.

WE guarantee that our Standard 2000 c. p. Light can be produced on .75 H. P. each, and that it will successfully compete with any 2000 c. p. light in the world for Size, Color and Steadiness. We guarantee to produce a light one-third larger than standard 1200 c. p. lights on same power that said 1200 c. p. lights require.

REGULATION.

WE guarantee Instantaneous Automatic Regulation so perfect that lights can be turned out and power reduced, and that one light can be maintained on our largest dynamo 24 hours or longer without heating the machine.

SEND FOR CATALOGUE.

The Waterhouse Electric & Mfg. Co.
HARTFORD, CONN.

THE
Baldwin Gas Engine

The Simplest, most Efficient, and Steadily Running Gas Engine ever built.

Adapted for

ELECTRIC LIGHTING

And all Industrial Purposes.

Otis Brothers & Co.

Elevators and Hoisting Machinery,

38 PARK ROW, - NEW YORK.

——THE——
BISHOP GUTTA PERCHA CO.

420 – 426 East 25th St., N. Y.,

MAKE A SPECIALTY OF

HIGH INSULATION

FOR EVERY ELECTRICAL PURPOSE.

WE have the oldest establishment in the U. S., and the best experts for this business. We do not make any kind of **Undertaker's Wire**, but thoroughly **Water-Proof, Fire-Proof, Acid-Proof** and **Alkili-Proof** Wires and Cables. Our **Gutta Percha Sub Aqueous Cables, Balata Flexible Cords** and **India Rubber Insulation** for general use are unexcelled. Tell us your difficulties and we will try to overcome them.

HENRY A. REED, Secretary and Manager.

The Pond Engineering Company,

ENGINEERS AND CONTRACTORS OF

STEAM AND HYDRAULIC MACHINERY,

ARE PREPARED TO FURNISH AND ERECT

COMPLETE STEAM PLANTS for
ELECTRIC LIGHT and POWER,

ENGINES, BOILERS, FURNACES, GRATES, HEATERS, PUMPS, INJECTORS, PIPE WORK, VALVES, BELTING, Etc.

Also to Deliver and Erect same, including Foundations, Brickwork, Pipe Fitting, etc., the whole delivered to purchaser ready for service. This work is in charge of Experienced Engineers, and particular attention is paid to

ECONOMY OF FUEL.   **SIMPLICITY OF CONSTRUCTION.**   **EASE OF OPERATION.**

THE recent great Developments in the uses of Dynamo Electricity for Illumination and for Power, have led us to devote special attention to the Design and Erection of Steam Plants for this class of service. In order to be commercially successful, such a plant should include the most recent improvements, and be in accordance with the best engineering practice of the day. Having made a specialty of this class of work for years, we are prepared to guarantee satisfaction.

REFERENCES:

Exposition and Music Hall Association, St. Louis.		Vincennes El. L. & Power Co.,	Vincennes, Ind.
J. Kennard & Sons Carpet Co.,	"	"Times" Building,	Kansas City, Mo.
Public School Library,	"	Dime Museum,	" "
St. Louis Sugar Refinery,	"	Grand Missouri Hotel,	" "
Thomson-Houston Electric Co.,	"	Metropolitan Railway Co.,	" "
Pope's Theatre,	"	Bona Venture Building,	" "
Grand Opera House,	"	Grand Ave. Cable Railway,	" "
"Globe Democrat,"	"	National Agricultural Exposition,	" "
"Westliche Post,"	"	Kansas City Cable Railway Co.,	" "
M. A. Seed Dry Plate Co.,	"	Warder Grand Opera House,	" "
The "Grand" Billiard Hall,	"	Thomson Houston El Light Co.,	St. Joseph, "
U. S. Steamer "Mississippi,"	"	Union Railway Co.,	" "
N. K. Fairbank & Co.,	"	State Lunatic Asylum,	Fulton, "
Burrell, Comstock & Co.,	"	State Deaf and Dumb Institute,	" "
Hotel Beers,	"	State University,	Columbia, "
John Plate & Co., Electro Depos., South St. Louis.		Electric Light and Power Co.,	Mexico, "
University of Kansas,	Lawrence, Kan.	Nevada Electric Light Co.,	Nevada, "
Western Br. Soldiers' Home, Leavenworth,	"	Greely Electric Light Co.,	Greeley, Colo.
Kansas Penitentiary,	Lansing, "	Holden Smelting Co.,	Denver, "
Kansas Insane Asylum,	Topeka, "	Electric, Gas Light and Fuel Co.,	Laramie, Wyo.
Edison Illuminating Co.,	"	Davenport Gas Light Co.,	Davenport, Iowa.
Water, Light and Tel. Co.,	Hutchinson, "	Capital City Electric Light Co.,	Des Moines, "
Great Bend Electric L. Co.,	Great Bend, "	Oskaloosa Gas Light Co.,	Oskaloosa, "
Abilene Water and El. L. Co.,	Abilene, "	Ottumwa Water Works,	Ottumwa, "
Parsons Light and Heat Co.,	Parsons, "	Austin Water, L. & Power Co.,	Austin, Tex.
Mendota Electric Light Co.,	Mendota, Ills.	Edison Illuminating Co.,	Palestine, "
Western Nail Co.,	Belleville, "	Paris Gas and Electric Light Co.,	Paris, "
Jerseyville El. L., Gas & Power Co.,	Jerseyville, "	Water and Electric L. Co.,	Nebraska City, Neb.
Olney Edison Electric Light Co.,	Olney, "	Gas and Electric Light Co.,	Fremont, "
Jenney El. L. and Power Co.,			Fort Wayne, Ind.

CORRESPONDENCE SOLICITED.

POND ENGINEERING COMPANY
707 and 709 Market Street, ST. LOUIS, Mo.

BRANCH: Room 31, Water Works Building, 600 Walnut Street, KANSAS CITY, Mo.

CHAS. K. WEAD, President.   H. M. LINNELL, General Manager.

THE HARTFORD DYNAMIC CO.

CONTRACTORS AND EXPERTS
FOR ELECTRICAL AND STEAM PLANTS

THE RUSSELL ENGINE

GIVES THE BEST SATISFACTION.

PLANS, SPECIFICATIONS,
AND
EXPERT ADVICE
ON
ELECTRIC LIGHTING
Furnished For
CITY COUNCILS
OR LOCAL COMPANIES.

253 Main St. HARTFORD, CONN. 253 Main St.

UNDERGROUND
Electric • Light • Wires
ARC or INCANDESCENT,

Solid or Drawing in System, as Laid in New Brunswick, N. J., Cincinnati, O., Brockton, Mass., Detroit, Mich., Boston, Mass., Columbus, Ohio., San Francisco, Cal., &c., &c.

BITITE WIRES for Damp places.
TRINIDAD WIRES for Pole or Overhead lines.

SEND FOR SAMPLES AND PRICES.

CALLENDER INSULATING & WATERPROOFING CO.
45 BROADWAY, N. Y.
CHICAGO.   BOSTON.   LONDON, England.   BRUSSELS, Belgium.

W. B. PEARSON,
Mechanical Engineer & Contractor
Room 436, "The Rookery,"
CHICAGO.

Makes a specialty of the Installation of Complete Steam Plants, in connection with the Engines manufactured by A. L. Ide & Son, Springfield, Ill. These Engines are designed especially for Electric Lighting Service, and are recommended by all Electric Companies. Estimates cheerfully furnished. Correspondence solicited. Parties visiting the city on business connected with Electric Lighting will find my office centrally located, and are invited to make it headquarters. It is in the same building with the U. S. Co., across the street from the Edison Co., and within five minutes' walk of the T.-H. Co., the Excelsior Co., and the Brush Co.

H. G. CHENEY, Pres.   M. S. CHAPMAN, Vice-Pres.   ROBERT CHENEY, Sec. and Treas.
N. T. PULSIFER, Gen'l Manager.   WM. A ANTHONY, Consulting Electrician.

THE
Mather Electric Co.

Manchester, Conn.

New York Office, 35 Broadway.
Chicago Office, 38 La Salle St.
Boston Office, 106 Summer St.
Cincinnati Office, Carlisle Building.

The Mather System for Incandescent Lighting

THE Dynamos of this system are equal to any in efficiency. The lamps are adapted to a higher potential than those of any other system, have no superior in life or efficiency, and

Do Not Blacken in Use.

Our Dynamo is very simple in construction, runs without sparks at the commutators and can be operated by any one accustomed to the care of engines or machinery. The system has no equal for mills, machine shops, or other places requiring isolated plants. We shall be pleased to furnish estimates for complete installations, with or without motive power, and will send one of our experts to examine the premises, if necessary.

The Electrical Construction Co.

IF you want any Electrical work done, write us and get our prices and catalogue. We are furnishing the best residences built with all kinds of Electrical Appliances. We make a specialty of wiring new buildings for Incandescent lights. Contracts taken for erecting Electric Light Plants complete.

HOLMES, BOOTH & HAYDENS,
25 PARK PLACE, NEW YORK.
—MANUFACTURERS OF—
BARE and INSULATED WIRE.

Underwriters' Copper Electric Light Line Wire, handsomely finished, highest conductivity.
Copper Magnet Wire, Flexible Silk and Worsted Cords for Incandescent Lighting.
Insulated Iron Pressure Wire. Patented Barbed Copper Lighting Protector Wire.
Lead Covered Copper Wire, for inside use.
Round and Flat Copper Bars. for station work.

AGENTS FOR
SOLID CARBONS FOR ELECTRIC LIGHTING.

PATENT "K. K." LINE WIRE,
FOR ELECTRIC LIGHTING,
TELEGRAPH AND TELEPHONE.

J. L. BARCLAY, Selling Agent, 185 Dearborn Street, Chicago, Ill.
THOS. L. SCOVILLE, New York Agent.

FACTORIES: Waterbury. Conn.

THE "CLARK" WIRE

INSULATION GUARANTEED WHEREVER USED, AERIAL, UNDERGROUND OR SUBMARINE.

In a letter from the Inspector of the Boston Fire Underwriters' Union, under date of March 29, 1886, he says: "A Thoroughly Reliable and Desirable Wire in Every Respect."

The rubber used in insulating our wires and cables is specially chemically prepared, and is *guaranteed to be waterproof*, and *will not deteriorate, oxidize or crack*, and will remain flexible in extreme cold weather, and not affected by heat. The insulation is protected from mechanical injury by one or more braids and the whole slicked with Clark's Patent Compound, which is water, oil, acid and, to a very great extent, fire-proof. *Our insulation will prove durable when all others fail.* We are prepared to furnish Single Wires of all gauges and diameter of insulation for Telegraph, Telephone and Electric Lights from stock. Cables made to order.

Eastern Electric Cable Comp'y
61 and 63 Hampshire St., Boston, Mass.

HENRY A. CLARK, General Manager.   HERBERT H. EUSTIS, Electrician.

THE
ECLIPSE
FRICTION
CLUTCH

NOW doing Heaviest Service in the Largest ELECTRIC LIGHT PLANTS. Sold on its Merits. Send for Catalogue.

Eclipse Wind Engine Co.
BELOIT, WIS.

The Parker-Russell Mining and Mfg. Co.

ADAPTED TO ALL SYSTEMS.

Uniformity,
Light and Life.

CITY OFFICE:
711 Pine St., ST. LOUIS, MO.

THE
Fort Wayne "Jenney" Electric Light Co.,

FORT WAYNE, IND.,

MANUFACTURERS OF

Electric Lighting Apparatus.

Our Dynamos are Simple, Compact and Durable.
Our Lamps are perfectly Automatic
and Self-Regulating.

WE MANUFACTURE AND CONTROL THE

✢ Slattery Induction System ✢

The most Scientific and Complete Alternating
System in existence.

Armatures ⊛ Converters Guaranteed.

For Catalogues and Estimates address

Fort Wayne "Jenney" Electric Light Co.,

General Office and Works, Fort Wayne, Ind.

New York Office: 242 and 244 E. 122d St., New York Electric Construction Co.
Chicago Office: 225 Dearborn St., W. J. Buckley, Manager.
Philadelphia Office: 26 N. 7th St., G. A. Wilbur, Manager.
Mexico Office: Mexico City, F. Adam, Successors.

ANSONIA BRASS & COPPER CO.

Sole Manufacturers of COWLES' PATENTED

FIRE-PROOF AND WEATHER-PROOF
Electric Light Line Wire

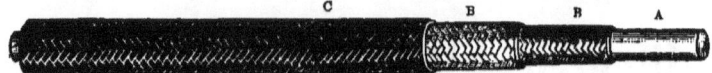

CUT SHOWING STYLE OF INSULATION.

A.—Copper wire.
B. B.—Two Braids saturated with Fire-Proof Insulation.
C.—Braided Cotton, saturated with a *Black*, WEATHER-PROOF Composition.

Samples furnished upon application. Pure Electric Copper Wire, Bare and Covered, of Every Description.

Warerooms, 19 & 21 Cliff St., New York.   64 Washington St., Chicago, Ill.

FACTORIES, ANSONIA, CONN.

SAWYER-MAN ELECTRIC CO.

Commercial Agent of

THE CONSOLIDATED ELECTRIC LIGHT CO.

The Best Lamp,
The Best Dynamo,
The Best Installation

The Best System of Isolated Incandescent Electric Lighting in the Market.

General Offices and Factory,   510-534 West 23d Street,

NEW YORK.

ESTABLISHED 1820.

ALFRED F. MOORE,

Insulated Electric Wire, Flexible Cords and Cables.

200 & 202 N. Third St. and 301 & 303 Race St.,

PHILADELPHIA, PA.

Single, Double, Triple and Quadruple Covered Wire, for Armatures and Field Magnets of Dynamo-Electric Machines and Motors.
Flat, Square or Rectangular Wire, Single or Double Covered.
Silk and Cotton Magnet Wire, Single or Double Covered.
Electric Light Line Wire.
Underwriters' Wire.   Weatherproof Wire.
Lead-Encased Electric Light Wire.
Flexible Electric Light Wire.
Lead-Encased Wires and Cables.
Anti-Induction Telephone Cables.
Office and Tower Cables.   Rubber Tape Cables.
Overhead, Underground and Sub-Marine Cables.
Office Wire, Single and Double Conductor.
Annunciator and Burglar Alarm Wires, Single, Double and Triple Covered.
Gas Fixture Wire, Triple Covered.
Gas Fixture Wire, Inner Wrap Silk.
Lead-Encased Wires, for Burglar Alarms, Call Bells and Gas Lighting.
Leading and Connecting Wire and Cable, for Blasting.

German Silver Resistance Wire, Single or Double Covered; Silk or Cotton.
Bare Copper and German Silver Wire.
Swedish and Charcoal Iron Wire, Bright or Tinned, Bare or Covered, for Motors.
Double Braided Galvanized Iron Line Wire, Weatherproof or Plain Insulation.
Hard-Drawn Copper Underwriters' Wire.
Hard-Drawn Copper Weatherproof Wire.
Insulated Pressure Wire.
Phosphor Bronze, Hard Copper and Steel Wire. Bare and Insulated, for Acoustic Telephones
Electrolier Wire.
Flexible Electric Cordage.
Incandescent Lamp Cord, Balata Insulation and Rubber Insulation.
Elevator Signal Cables and Lamp Cables.
Arc Lamp and Shunt Cords.
Telephone Cords and Switch Cords.
Push Button, Pole and Battery Cords.
Twisted Tinsel Cord and Wire Tinsel Cord, Silk and Cotton Covered, for Medical Batteries.
Dental and Motor Cords, &c., &c., &c.

American Conduit & Construction Co.

—MANUFACTURERS OF—

CARBONIZED STONE CONDUITS

FOR ELECTRIC WIRES.

14½ State Street, Room 42,   -   -   BOSTON, MASS.

FACTORY, NEPONSET AVENUE, WARD 23.

COMPLETE STEAM PLANTS UNDER ONE RESPONSIBILITY.

The Jarvis Engineering Co.

BOSTON,   **NEW YORK,**   **CHICAGO,**
61 Oliver St.   109 Liberty St.   81 Lake St.

MAKE A SPECIALTY OF

STEAM PLANTS

FOR

Electric Lighting and Power Stations

COMPRISING

ARMINGTON AND SIMS' ENGINES,
JARVIS FURNACES, Etc.

ENSURING

Economical Production of Power,
Steadiness of Lights,
Durability of Plant.

Send for Circulars and References.

The National Feed-Water Heater.

Over 100,000 Horse Power

IN USE IN THE UNITED STATES.

❦) **200** (❧

Electric Light Stations SUPPLIED.

PRICES LOW.

SATISFACTION UNIVERSAL.

Seventeen sizes manufactur'd, 6 to 2,000 Horse Power Capacity.

Heats the Water for the Boilers up to 206° to 212° F.

Send for List of Users and Examine for Yourself.

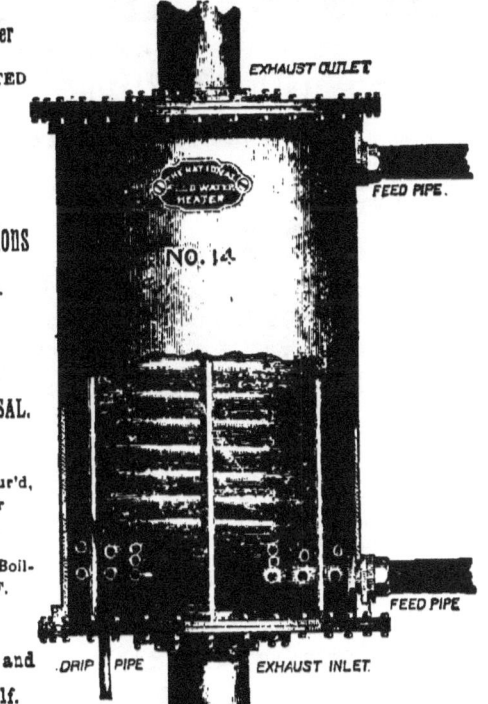

THE NATIONAL PIPE BENDING CO.
84 River Street, NEW HAVEN, CONN.

CLEVERLY ELECTRICAL WORKS.

1018 Chestnut Street, PHILADELPHIA.

ELECTRICAL SUPPLY MAKERS.

THE
Hussey Re-Heater
——AND——
STEAM PLANT IMPROVEMENT CO.

15 CORTLANDT STREET,

A. S. HATCH, President.
S. D. BREWER, General Manager.
LEVI HUSSEY, Engineer.

NEW YORK.

Consulting and Practical Experts in all matters pertaining to Steam, and its application to Power, Heat and Ventilation.

Designing, Remodeling and Improving Steam Plants for Office Buildings, Stores, Apartment Houses, Hotels, Manufacturing Establishments, and

Electric Light Companies.

SOLE PROPRIETORS OF THE

Hussey Re-Heater System
FOR

Re-Heating Exhaust Steam,
    Super-Heating Live Steam,
        and Heating Air and Water,
            Without Cost for Fuel,

BY THE USE OF WHICH

An Economy of from 25 to 50 per cent, with Increased Efficiency at the Reduced Cost, can be ensured in any Steam Plant to which it is applied; and the

Exhaust Steam of Electric Light Plants can be Converted into an Important Source of REVENUE.

The E. S. GREEELY & CO.

Successors to L. G. TILLOTSON & CO.,

IMPORTERS. **Electric Light** **MANUFACTURERS.**
SUPPLIES

5 and 7 DEY STREET, NEW YORK.

AYRTON & PERRY'S VOLT and AM-METERS.

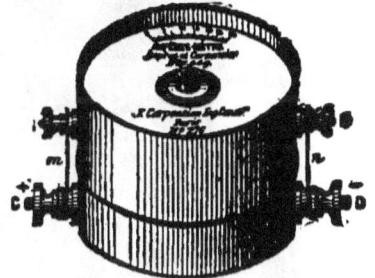

CARPENTIER'S VOLT and AM-METERS.

LINE Material, Station Equipments, Construction Tools. Everything that is necessary for Installations and Maintenance kept constantly in stock. Telegraph, Telephone and General Electrical Supplies. Incandescent Lamps for Battery Use. Wholesale Agents for Fletcher's Gem Wire Holders and Sleet-Proof Pulleys, Cleveland's Gang Switches and Arc Light Cut-Outs. Scientific Electrical Measurement Apparatus. General Agents for the Standard Electrical Test Instruments of the Electric Manufacturing Co., of Troy.

American Leather Link Belt Co.

A New Article

MADE OF SMALL

LEATHER

LINKS.

SPECIALLY ADAPTED FOR USE ON

DYNAMOS

ENDORSED BY

ALL PROMINENT ELECTRICIANS

Chas. A. Schieren & Co.

MANUFACTURERS AND SOLE AGENTS.

86 Federal Street, BOSTON.
47-61 Ferry Street, NEW YORK.
46 South Canal Street, CHICAGO.
416 Arch Street, PHILADELPHIA.

The AMERICAN TOOL AND MACHINE CO.

ENGINEERS, FOUNDERS AND MACHINISTS.
84 Kingston St., Boston, Mass.

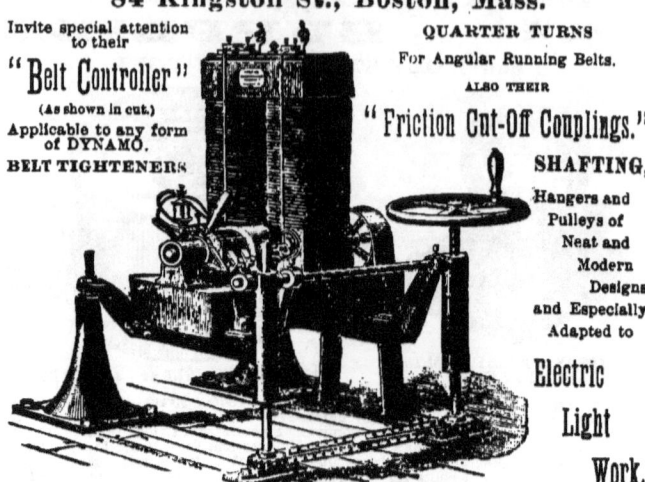

Invite special attention to their

"Belt Controller"

(As shown in cut.)

Applicable to any form of DYNAMO.

BELT TIGHTENERS

QUARTER TURNS

For Angular Running Belts.

ALSO THEIR

"Friction Cut-Off Couplings."

SHAFTING,

Hangers and Pulleys of Neat and Modern Designs and Especially Adapted to

Electric Light Work.

For Prices, Drawings, &c., address the Company at their office, 84 Kingston Street, Boston, Mass.

The Schuyler Electric Co.

MANUFACTURERS OF

DYNAMO MACHINES,
ARC AND INCANDESCENT LAMPS.

This System has the Most Perfect

AUTOMATIC REGULATOR

ON THE MARKET.

Write for Illustrated Circulars and Estimates.

THE SCHUYLER ELECTRIC CO.
MIDDLETOWN, CONN.

MAIN BELTING COMPANY,
PHILADELPHIA.   CHICAGO.

MANUFACTURERS OF THE

Leviathan Belting.

The Best Belt for Dynamos, Motors, and all Electrical Purposes.

UNEQUALLED FOR

Traction-Power,

Uniformity and

Durability.

MADE ANY LENGTH.

Widths, 1 to 90 inches.

BELTS MADE ENDLESS.

Strongest and Cheapest

IN THE WORLD.

Correspondence solicited. Please write for Prices and Samples,

Main Belting Company

Ninth and Reed Streets,
PHILADELPHIA.

248 Randolph Street,
CHICAGO.

At a meeting of the special committee appointed by the Common Council to examine and report upon the feasibility of the city owning and operating its own electric lighting plant, the following resolution was offered:

By Alderman Burt:

Resolved, That the thanks of this committee be, and they are hereby tendered to Mr. Fred. H. Whipple, secretary of the committee, for the very able, exhaustive and comprehensive manner in which he has discharged the laborious duties assigned to him, in preparing and compiling the information and facts connected with the subject relegated to the committee. We consider his work, as completed, to be beyond all question the most valuable digest of the matter that has ever yet been produced, and cheerfully recommend it to all who are interested in the subject of municipal lighting.

Adopted as follows:

Ayes—Ald. Burt, Meier, Trombley, Holihan, Amos.
Nays—None.

Detroit, June 9, 1888.

FIAT LUX ET ERAT LUX.

MUNICIPAL

LIGHTING

By Fred H. Whipple.

Entered according to the Act of Congress, in the year 1888, in the Office of the Librarian of Congress, at Washington, D. C.

DETROIT, MICH.
1888.

To Hon. M. H. CHAMBERLAIN, of Detroit, who, when in the Executive Chair did so much to further an investigation of the Public Lighting, this little volume is dedicated by the writer.

CONTENTS.

	PAGES
Advertisements...	1-25
Introductory..	33-34
The Electric Light...	35-37
The Contract or Rental System.............................	38-57
Municipal Lighting...	58-77
How to Buy a Plant..	78-90
Economy in Steam...	91-93
Importance of Belting.....................................	94-97
How to Light a City.......................................	98-104
The Different Systems—	
American...	105-108
Ball..	108
Brush..	108-112
Clark...	112-114
Edison...	143-144
Excelsior..	114-116
Fort Wayne Jenney...................................	116-119
Heisler...	119-121
Hill..	121
Indianapolis Jenney..................................	123-126
Loomis...	126
Mather...	127
Mutual...	128-129
Sawyer-Man..	130
Schuyler...	131
Thomson-Houston....................................	132-134
United States..	134
Van Depoele...	135-138
Waterhouse..	138-141
Western..	141-142
Westinghouse..	144-146
Storage Batteries..	147-149
Distribution of Light......................................	150-169
Underground Lines..	170-226
Addenda...	227
Advertisements..	228-257

INDEX.

ILLUSTRATIONS

Adams Tower, 156.
American Dynamo, 106.
American Street Lamp, 107.
Brush Dynamo, 109.
Brush Arc Lamp, 110.
Brush-Swan Lamp, 111.
Clark Dynamo, 113.
Clark Lamp, 114.
Detroit Tower, 161.
Edison Dynamo, 142.
Edison Lamp, 143.
Excelsior Dynamo, 115.
Excelsior Lamp, 116.
Electric Leather Belting, 95.
Fort Wayne Jenney Dynamo, 117.
Fort Wayne Jenney Lamp, 118.
Heisler Dynamo, 119.
Heisler Lamp, 120.
Hide of Belt Leather, 94.
Hill Dynamo, 121.
Hill Lamp, 122.
Incandescent Street Lamp, 168.
Indianapolis Jenney Dynamo, 124.
Indianapolis Jenney Lamp, 125.
Indianapolis Tower, 160.
Lamp Hangers, 155.
Loomis Dynamo, 126.
Loomis Lamp, 127.
Mast Arms, 153.
Mather Dynamo, 127.

Mather Lamp, 127.
Mutual Dynamo, 128.
Mutual Lamp, 129.
Patent Joint Belt, 96.
Patent Joint Belt in Operation, 96.
Plan of Intersection Lighting, 154.
Sawyer-Man Lamp, 129.
Sawyer-Man Dynamo, 130.
Schuyler Dynamo, 131.
Schuyler Lamp, 132.
Star Iron Tower, 159.
Storage Battery, 149.
Street Pole Light, 152.
Thomson-Houston Dynamo, 133.
Thomson-Houston Lamp, 134.
United States Dynamo, 135.
United States Incandescent Lamp, 135.
United States Arc Lamp, 136.
Van Depoele Dynamo, 137.
Van Depoele Incandescent Lamp, 138.
Van Depoele Arc Lamp, 138.
Waterhouse Dynamo, 139.
Waterhouse Lamp, 140.
Western Lamp, 140.
Western Dynamo, 141.
Westinghouse Street Light, 145.
Westinghouse Street Light, 146.
Westinghouse Converter, 146.

SYSTEMS.

American, 61, 65, 105, 106, 107.
Ball, 108.
Brush, 65, 66, 68, 70, 75, 76, 108, 182, 183, 184, 186, 190, 204, 210, 217, 221, 227.
Clark, 5, 112, 113.
Edison, 142, 180, 194, 201, 203, 227.
Excelsior, 114, 179.
Fort Wayne Jenney, 62, 65, 67, 75, 116, 117, 118, 179.
Heisler, 119.
Hill, 121.
Indianapolis Jenney, 59, 64, 65, 66, 69, 70, 123, 124, 125, 218, 227.
Loomis, 126.

Mather, 13, 127.
Mutual, 128.
Sawyer-Man, 129.
Schuyler, 65, 131.
Sperry, 179.
Thomson-Houston, 7, 65, 66, 68, 69, 70, 132, 133, 179, 182, 183, 194, 203, 209, 210, 216.
United States, 1, 134, 135, 136, 182, 186, 190, 227.
Van Depoele, 4, 75, 135.
Waterhouse, 9, 138.
Western, 6d, 69, 71, 74, 76, 140, 179.
Westinghouse, 144, 145, 146.

LOCALITIES.

Adrian, Mich., 38.
Akron, Ohio, 38, 164.
Albany, N. Y., 38.
Albuquerque, N. M., 38.
Allentown, Pa., 39.
Alliance, Ohio, 38.
Asheville, N. C., 39.
Ashland, Pa., 39.
Attica, Ind., 227.
Atlanta, Ga., 39.
Augusta, Ga., 39.
Aurora, Ill., 65, 68, 76.
Baltimore, Md., 39, 205, 207.
Bangor, Me., 39.
Bath, Me., 39.
Battle Creek, Mich., 39.
Bay City, Mich., 59, 60, 65, 76, 165.
Bloomington, Ill., 39.
Boston, Mass., 40, 121, 201, 204, 208.
Brockton, Mass., 203, 227.
Brooklyn, N. Y., 40, 191, 204, 207, 208.
Bridgeport, Conn., 40.
Buffalo, N. Y., 40, 203.
Burlington, Ia., 40.
Burlington, Vt., 40.
Cambridge, Mass., 40.
Camden, N. J., 208.
Cedar Rapids, Ia., 40.
Champaign, Ill., 66, 76.
Charleston, S. C., 41.
Chattanooga, Tenn., 41.
Chicago, Ill., 67, 69, 71, 77, 93, 176, 177, 204, 207, 208, 209, 210, 213, 214, 217.
Chillicothe, Ohio, 41.
Cleveland, Ohio, 41, 210.
Columbus, Ga., 41.
Columbus, Ohio, 41.
Columbus, Ind., 65.
Concord, N. H., 41.
Conshohocken, Pa., 227.
Council Bluffs, Ia., 41, 164.
Danbury, Conn., 161.
Danville, Ill., 165.
Davenport, Ia., 42.
Dayton, Ohio, 42.
Decatur, Ill., 67, 166.
Defiance, Ohio, 42.
Denver, Col., 42, 163, 204, 209.
Des Moines, Ia., 42, 93.
Detroit, Mich., 57, 76, 77. 98, 157, 159, 161, 203, 209, 210, 213, 215.
Easton, Pa., 74.
East Liverpool, Ohio, 42.
East Portland, Ore., 121.

Eau Claire, Wis., 42.
Elgin, Ill., 42, 163.
Erie, Pa., 42.
Eugene City, Ore., 121.
Evansville, Ind., 42, 165.
Fairfield, Ia., 75, 167.
Fall River, Mass., 43.
Fargo, Dak., 43, 163.
Fitchburg, Mass., 43
Flint, Mich., 162.
Fond Du Lac, Wis., 43, 163.
Fort Wayne, Ind., 43, 167.
Frederickton, Md., 227.
Galesburg, Ill., 43.
Galveston, Tex., 43.
Gloucester, Mass., 43.
Goshen, Ind., 166.
Grand Ledge, Mich., 70.
Hannibal, Mo., 62, 76.
Harrisburg, Pa., 44.
Hartford, Conn., 44.
Haverhill, Mass., 44.
Hoboken, N. J., 44.
Holyoke, Mass., 44.
Hornellsville, N. Y., 44.
Houston, Tex., 44.
Huntington, Ind., 67, 76.
Indianapolis, 65, 160.
Jackson, Mich., 44.
Jacksonville, Ill., 44, 167.
Janesville, Wis., 45.
Jersey City, N. J., 45.
Joliet, Ill., 45.
Kalamazoo, Mich., 45.
Kansas City, Mo, 45.
Keene, N. H., 45.
Keokuk, Ia., 45.
La Crosse, Wis., 45, 166.
Lafayette, Ind, 45.
Lancaster, Pa., 45.
Lawrence, Mass., 46.
Lewiston, Me., 61, 76.
Liberty, Mo., 46, 121.
Lima, Ohio, 46.
Little Rock, Ark., 75.
Lockport, N. Y., 46.
Logansport, Ind., 46.
London, Ont., 46.
Lowell, Mass., 46.
Lynn, Mass., 46.
Lyons, Ia., 74.
Macon, Ga., 47, 166.
Madison, Ind., 64, 65, 76.
Manchester, N. H., 47.
Mankato, Minn., 121.
Mansfield, Ohio, 47.

Martinsville, Ind., 68.
Massillon, Ohio, 47.
Matteawan, N. Y., 47, 121.
Memphis, Tenn., 47.
Michigan City, Ind., 60, 77.
Milwaukee, Wis., 47, 209.
Mobile, Ala., 47.
Monmouth, Ill., 227.
Montgomery. Ala., 47.
Monticello, Minn., 47, 121.
Montreal, Canada, 48, 93.
Nashville, Tenn., 48.
Newark, N. J., 48.
New Bedford, Mass., 48.
New Britain, Conn., 48.
Newburgh, N. Y., 48.
New Haven, Conn., 48.
New Orleans, La., 48.
Newton, Mass. 48.
New York, 49, 186, 196, 204, 207, 208, 209, 210, 213, 216, 225, 227.
Norfolk, Va., 49.
Northampton, Mass., 49, 76.
Norwalk, Ohio, 49.
Ocean Grove, N. J., 121.
Ogden, Utah, 49, 163.
Olney, Ill., 227.
Omaha, Neb., 49.
Orange, N. J., 49.
Oskaloosa, Ia., 227.
Oswego, N. Y., 50.
Ottawa, Canada, 50.
Ottawa, Kan., 50.
Owego, N. Y., 113.
Painesville, Ohio, 69.
Paris, Ill., 62, 63, 76.
Paterson, N. J., 50.
Peekskill, N. Y., 227.
Pendleton, Ore., 121, 227.
Peoria, Ill., 50.
Peru, Ill., 227.
Petersburg, Va., 50.
Philadelphia, Pa., 50, 57, 70, 181, 196, 204, 206, 207, 208, 209, 210, 215, 216, 217, 218, 225.
Pittsburg, Pa., 51, 93, 185, 204, 209, 215, 217, 218, 222.
Plainfield, N. J., 227.
Portland, Me., 51.
Portland, Ore., 51.
Portsmouth, N. H., 51.
Portsmouth, Ohio, 69, 77.
Pottsville, Pa., 51.
Poughkeepsie, N. Y., 51.
Providence, R. I., 51.
Quincy, Ill., 51.
Racine, Wis., 51.
Reading, Pa., 52.
Red Bank, N. J., 121.
Richmond, Va., 52.
Rochester, N. Y., 52.

Rock Island, Ill., 52, 166.
Rome, N. Y., 52.
Sacramento, Cal., 52.
Saginaw, Mich., 52, 166.
Salem, Mass., 53.
Salem, Ohio, 227.
San Antonio, Tex., 53.
Sandusky, Ohio, 53.
San Francisco, Cal., 53.
Savannah, Ga., 53.
Schenectady, N. Y., 53.
Scranton, Pa., 53.
Seattle, W. T., 227.
Sedalia, Mo., 54.
Selma, Ala., 54.
Sherman, Tex., 75.
Somerville, Mass., 54.
South Bend, Ind., 54.
Springfield, Mass., 54, 185, 209, 222.
Springfield, Ohio, 54.
St. Joseph, Mo., 52.
St. Louis, Mo., 65.
Stillwater, Minn., 54.
Stockton, Cal., 54.
Syracuse, N. Y., 54.
Tacoma, W. T., 227.
Taunton, Mass., 54.
Terre Haute, Ind., 55, 65.
Tipton, Ia., 165.
Toledo, Ohio, 55.
Topeka, Kan., 64, 65, 66, 76.
Toronto, Ont., 55.
Torrington, Conn., 227.
Trenton, N. J., 55.
Troy, N. Y., 55.
Tyler, Tex., 227.
Tyrone, Pa., 227.
Union City, Ind., 55.
Urbana, Ohio, 227.
Utica, N. Y., 55, 167.
Vicksburg, Miss., 56.
Vincennes, Ind., 120.
Wabash, Ind., 56, 121.
Waltham, Mass., 56.
Washington, D. C., 56, 57, 190, 204, 205, 207, 208, 210.
Waterbury, Conn., 56.
Watertown, N. Y., 56.
Wheeling, W. Va., 227.
Wichita, Kan., 56, 65.
Wilkesbarre, Pa., 56.
Williamsport, Pa., 56.
Wilmington, Del., 208.
Winona, Minn., 57.
Woburn, Mass., 57.
Wooster, Ohio, 57.
Worcester, Mass., 57.
Yonkers, N. Y., 57.
Youngstown, Ohio, 57.
Ypsilanti, Mich., 69, 77.

Introductory.

THE writer of the appended pages had no thought of writing a book until it was written. The ideas that have culminated in this volume were developed through natural causes, and sprung from sources which in themselves had no connection with book-making.

The electric light, as a scientific curiosity, is old, but it is little understood by those who are the largest patrons of its benefits. No attempt has been made to treat of it except so far as to be understood by electricians. The great public has never been told in language stripped of its technicalities, of the real value of electricity as an illuminant. Municipal and private corporations and capital seeking investment know of the electric light only as it shines upon the streets, or as they read of it in the pamphlets of the manufacturing companies. The people have never had an inning; the peoples' representatives know little about their public lighting, and, in some cases perhaps, care less. For this they are excusable. The subject as it has been presented to them has been too deep for easy solution, and they have paid for their light, not on a basis of what they could buy it for, but on the basis of what the company could get. No effort to systematically obtain and arrange information upon the politico-economic questions has heretofore been made, and the contest between the supply and the demand has been necessarily one sided.

The title of this volume is meant to be definite. It is not the design of the writer to dip into the scientific questions that surround the subject of electric lighting, nor to compare the values of different systems and makes. These matters are more pertinently in the hands of those who have light to sell and those who want to buy. The object of this volume is simply to contribute a mite to the general knowledge of the present status of municipal electric lighting, and to indicate, so far as possible, what should be done by the public treasury or the private purse in investing in electric lighting.

To this end the writer has sought to obtain information upon such points as the laymen will desire to know. With what success this has been accomplished the reader may judge. One merit, if no other, this little volume contains—the information is authentic, its presentation is impartial, and its resumé is thorough. To municipal bodies, therefore, this volume is in particular intended, and in general to all who have any interest in the subject of electric lighting.

The Electric Light.

URING the period extending from the introduction of the telegraph to the invention of the telephone, the efforts of the electrical world were devoted mainly to the development of Prof. Morse's immortal discovery. Indeed, the telephone itself —one form of it at least—was the outgrowth of observations and experimental investigations, having for their object the improvement of the telegraph. The success of the telephone, which was almost electric, caused an awakening in electrical science. Facts that had long been known, but which were esteemed merely amusing experiments, were studied with renewed interest, and the results were the brilliant achievements which so startled the world a few years back. These were, notably, the electric light and the electric railway.

The history of the electric light is interesting. Notwithstanding the proverbial tendency of mankind to believe that the "olden time" showed many points of superiority to the present, the casual retrospective view will call to mind a list of modern improvements which are now considered indispensable adjuncts of civilization, but which have been introduced within the memory of thousands who are now living. These inventions, like that which is the special subject of these pages, instead of being born full-grown, were more or less imperfect in their infancy, and have been and are still subjects of constant experiments and improvements. The records of the patent office illustrate the number of men whose active brains are employed in the constant search for something better than the world has yet seen.

In the days of the hand-loom and the spinning-wheel, of the stage coach and the sailing ship, our forefathers extended their hours of labor by the aid of the pine knot and the tallow dip, and sought in vain for more efficient means of illumination until the hardy whalemen of New Bedford and Nantucket provided the malodorous "whale oil," whose advent was thought to mark an era in the world's advancement.

One can easily recall the excitement attendant upon the discovery of petroleum in Pennsylvania, and the successive steps which resulted in the now almost universal kerosene lamp, whose odor, inconvenience, and proneness to cause destructive conflagrations, are sufficiently familiar to cause it to be regarded as by no means an unmixed blessing.

The inventive genius of mankind next brought forward coal-gas,—a vast improvement on its predecessors, but handicapped by its own peculiar disadvantages. The cost of its introduction limits its field of usefulness to cities and large towns, and the capital invested in its production, and in the means of its conveyance to the consumer, requires the payment of heavy dividends, and, in connection with that potent factor, the conscienceless "meter," draws such sums from the user's pocket that he often returns to the despised kerosene lamp, or, unable to make that sacrifice of convenience and safety, waits with eager anxiety for the coming of some means of illumination as safe and convenient as gas, while cheaper and more reliable.

For many years it has been known that an extremely brilliant light could be produced by slightly separating two pencils of carbon, through which a powerful current of electricity was passing, as the mysterious force spans the gap with an "arc" of intense light. The one insuperable bar to the general introduction of this light, was its great cost, due to the necessity of producing the current by the consumption of zinc in the galvanic battery. In spite of this expense the arc light early found a limited application to lighthouses, and other important government works.

The solution of the problem was seen, by the prophetic minds of scientific men, to depend upon the discovery of an economical means of generating the electric current, and many years were spent in fruitless endeavors to reduce its cost. Finally, advantage was taken of the fact—long known as a scientific curiosity — that whenever a wire or other conductor is approached to or removed from a magnet, a current is produced in the conductor. This unused and almost forgotten bit of knowledge, unimportant as it seemed, was the key of the whole problem.

Given a means of generating the electric current, not by the consumption of costly chemicals, but by *power* — power derivable from any of the sources long utilized by man—and it was possible to produce economically the brilliant "arc" which now casts its intense white light on city streets the world over.

The first obstacle to the more general introduction of the electric light was, as stated, its cost, and in some quarters this has not entirely been done away with at the present day. For some purposes and in some places it has been impossible as yet to place electricity in competition with gas as regards expense. Wherever the comparison has been made electricity has won the day. But why draw a comparison? People do not compare the cost of gas with that of candles, nor the price of a pheasant with that of a mutton chop. People will have the electric light if it can be supplied to them, not because it is cheap, but because it is safe, healthy, pure, soft and natural. And, moreover, they will not object to paying a reasonable price for it, whatever may be the price of gas.

Hitherto the electric light has been regarded more as a luxury than as a light for general use. In 1870 there was not a single lamp lighted by electricity, and today there are in this country over 500,000 in operation. Is not, then, the topic of electric lighting a live one?

The Contract, or Rental System.

THE prices paid by American cities for public lighting, under the contract or rental system, together with the systems in use, the manner of stringing wires and hanging the lamps, the hours which the lamps are burned, the area lighted, and other information concerning their lighting contracts, will be found in the following pages. The data for the same is brought up to April 1st, 1888.

ADRIAN, Mich., has just contracted for 60 Thomson-Houston lights at $100 each per year, to burn all night and every night, the cost to be reduced when the number reaches 75. Three square miles will be lighted by the intersection system. The contract is for three years.

AKRON, Ohio, lights four square miles with 170 Thomson-Houston lights, swung at street intersections, burning up to 2 o'clock A. M., for 3 9-10 cents per hour per lamp, for 2,000 hours; contract, five years; wires overhead. In 1880 the city put up two towers and used nine Brush lights, of nominal 4,000 candle-power each. In 1884 the intersection system was adopted and the Thomson-Houston light, which, Newton Ford, City Clerk, writes, gives better general satisfaction.

ALBANY, N. Y., lights three square miles of its business territory with 481 Brush lights, placed on poles and at intersections. The city owns the poles, wires and lamps. The price of lights is 50 cents each per night, burning all night, or $182.50 per year. Contract, five years; wires overhead.

ALBUQUERQUE, New Mexico, has three Brush lights on its principal street placed on poles, and burning until 1 o'clock. They cost $192 each per year.

ALLIANCE, Ohio, pays $144 each per year for 80 Western

lights, which are placed at street intersections and burn all night. Contract, annual; wires overhead.

ALLENTOWN, Pa., has 100 American lamps, swung at street intersections, in four square miles, which burn all night, except on moonlight nights, and cost $100 each per year. Contract, five years; wires overhead. Naphtha is burned every night in the year for $21 per light.

ASHEVILLE, N. C., has 35 Jenney lights, placed on towers and poles, which burn until 2 o'clock, for $100 each per year. Two square miles are lighted. The contract expires this year.

ASHLAND, Pa., uses 23 Thomson-Houston lights at street intersections, lighting two square miles on the moon schedule, for $100 per light per year. The contract is for one year.

ATLANTA, Ga., uses 100 Thomson-Houston intersection all night and every night lights, which cost 32 cents each per night, or $120 per year; wires overhead; contract, three years. Gas and gasoline is also used.

AUGUSTA, GA., has but three lights, of the American system, in the public park. They burn until midnight and cost $100 each per annum. Gas and kerosene are used throughout the city.

BALTIMORE has no contract, but pays 50 cents per night per light, or $182.50 each per year, burning every and all night, for 519 Brush lights. The lamps are placed on poles and mast-arms; wires overhead. The lights are scattered in dark alleys and on principal streets.

BANGOR, Me., has only 23 lights of the Thomson-Houston system, which are used in the central parts, lighting one square mile. The lights burn all night, are placed on poles, with wires overhead, and cost $150 each per year. The contract is yearly.

BATH, Me., has 20 American lights burning, on poles, on the Philadelphia schedule, which cost $100 each per year. No other light is used. Contract, five years.

BATTLE CREEK, Mich., has 62 Thomson-Houston intersection lights, which light four square miles. No other light is used. Ten lamps burn all night; the remainder until midnight, and cost $111 each per year. Contract, one year.

BLOOMINGTON, Ill., pays $108 each per year for 211 intersection and pole Thomson-Houston lights, burning all night and all but moonlight nights; four square miles are lighted. Contract, three years; wires overhead.

BOSTON, Mass., places 570 Brush and Thomson-Houston lights on iron extension arms on regular lamp-posts. Twelve square miles are lighted under a three-years' contract, the lamps burning 3,828 hours per year, at a cost of 65 cents per night, or $237.25 per year. The wires are overhead, and the city is agitating placing them underground.

BRIDGEPORT, Conn., uses 94 Thomson-Houston lights, placed on poles and mast-arms, which burn all night and cost 50 cents each per night, or $182.50 per year. Four square miles are lighted under a five-years' contract. The wires are strung overhead on the telephone poles.

BROOKLYN, N. Y., covers part of the city with 1,007 Thomson-Houston lights placed on poles. The lamps burn 3,900 hours per year, and cost under the present annual contract 50 cents each per night, or $182.50 per year. The wires are strung overhead.

BUFFALO, N. Y., lights 45½ miles of street, using 631 Brush, Thomson-Houston and United States lights, which burn all and every night, at a cost of 47¼ cents each per night, or $174.38 per year. The lamps are swung on poles and at intersections, and the wires are overhead, but conduits, principally of creosoted wood, have been constructed for the Thomson-Houston wires. The contract is annual. The remainder of the city is lighted by gas.

BURLINGTON, Iowa, has 39 Van Depoele lights, placed on poles and at intersections. The lights are burned on the Philadelphia schedule, and cost $130 per lamp per annum. The contract is for three years. Gas is also used.

BURLINGTON, Vt., places 70 Brush lights on poles and mast-arms, which, with 50 naphtha lamps, light four square miles until midnight every night. The price under a three-years' contract is 32 cents per light per night, or $116 per year.

CAMBRIDGE, Mass., has 78 American lights on three of the main avenues at six miles of intersections. They burn all night and every night at 55 cents each per night, or $200 per year. There is no contract; wires overhead. The City Clerk reports an unsatisfactory condition of affairs.

CEDAR RAPIDS, Iowa, has forty-four Thomson-Houston lights

at street intersections, which light four square miles all night. The cost under a ten years' contract is $120 per lamp per year. The contract expires March 15, 1896.

CHARLESTON, S. C., has fifty Thomson-Houston all night pole lights, which cost $168 each per annum. Gas is also used. The contract expires January 1, 1889.

CHATTANOOGA, Tenn., pays 33⅓ cents each per night, or $121.66 per year, for 30 Brush lights, burning all night and every night. Poles and towers and overhead wires are used. Contract, two years.

CHILLICOTHE, Ohio, has 121 Brush lights placed on poles, which burn on the Philadelphia schedule. They cost, under a five-years' contract, $80 per lamp per year.

CLEVELAND, Ohio, lights but 1½ square miles with electric lamps, using the Brush light, 42 of 2,000 candle-power and 26 of 4,000 candle-power each. The former are placed on wires swung at street intersections, and cost 3 7-10 cents per hour each. The large lamps are on iron masts, and cost 10¼ cents each per hour. The lamps are burned 3,760 hours per year, making the cost for 2,000 candle-power $139.12 and 4,000 candle-power $394.80 each. Contract annual; wires overhead.

COLUMBUS, Ga., places 21 Brush lights on poles on two streets. They burn all night for $108 each per year, under a two-years' contract. The remainder of the city is lighted with gas.

COLUMBUS, Ohio, is paying for 200 Thomson-Houston lights, and by June will have 200 more. Intersection system is used exclusively with the arc light wires overhead. The Edison Company have their incandescent wires underground. The Philadelphia schedule is followed, and the cost is $48 per light per year for the first 50, and $80 per light for the others. Owing to the condition of the funds in the city there is no regular contract. Gasoline is used in the suburbs.

CONCORD, N. H., has gas and 17 Thomson-Houston pole lights, which burn until midnight and cost $100 per night per year. There is no contract.

COUNCIL BLUFFS, Iowa, has seven 150-foot towers of four Thomson-Houston lights each, for which it pays $240 per year for each light. The lights burn all night, except on moonlight nights. Wires overhead; contract, two years. One and one-half square miles are lighted.

DAVENPORT, Iowa, has a 20-years' contract for 54 mast-arm intersection and 40 tower lights of the Jenney system, which lights four and a half square miles when there is no moon. Tower lights cost $190 each per year, and low lights each $145 per year; wires overhead.

DAYTON, Ohio, pays $150 per year for each of the 126 Fuller arc lights used on the streets. The lights burn 3,674 hours per year; pole system; wires overhead; contract, three years. There are 965 gas-lamps, which cost $19 each per year.

DEFIANCE, Ohio, pays $60 per lamp per year for 52 Western intersection lights. Contract annual; wires overhead.

DENVER, Col., has 540 incandescent lamps on old gas posts and seven towers with four arc lights each. The Westinghouse alternating system is used with the wires underground. The price is $28 each per year for the incandescent and $855 per year for each tower. According to J. R. Treadway, City Clerk, the latter are not satisfactory and are being taken out, as the contract for them expires. The moonlight schedule for all night lighting is employed. Contract for the incandescent lamps, three years.

DES MOINES maintains six lights on her bridges, for which she pays $120 each per year, burning until midnight.

EAU CLAIRE, Wis., uses 48 Brush lights, placed on towers and at intersections, which are claimed to light twelve square miles. The lights burn all and every night, and cost $157 each per year. The contract expires this year.

EAST LIVERPOOL, Ohio, pays $85 each per year for its first 25 Western intersection lights, and $80 for the others. Contract, annual; wires overhead.

ELGIN, Ill., is lighted by 33 Van Depoele lamps placed on towers, which burn all night, and light four square miles, under a five-years' contract which expires this year. The cost is $7,000 per year, or $212 per lamp.

ERIE, Pa., uses 54 Brush lights, swung at intersections, and pays 40 cents each per night, or $146 per year. The lights burn all and every night, under an annual contract; wires overhead. One mile square is lighted.

EVANSVILLE, Ind., has ten towers of four lights each, 49

arches at intersections, and ten poles. The Brush system is used, and the lamps burn when the moon does not shine, at a cost of $24,480 per year, with a deduction when not burned of 10 cents per hour for each pole or arch, and 15 cents per hour for each tower light. Five square miles are lighted. The contract is for ten years. Overhead wires. No other light is used.

FALL RIVER, Mass., has 260 Thomson-Houston lights, distributed over the city, which burn every and all night, for $204 each per year. Poles are used with overhead wires; contract for one year.

FARGO, Dak., has three towers, with six Brush lights on each, and two poles. The lights burn all night, for $240 each per year. Contract, two years.

FITCHBURG, Mass., for six hours each night, is lighted by 51 Thomson-Houston lights, placed on poles, and also by gas and gasoline. The electric lights cost $90 each per year. Contract, annual.

FOND DU LAC, Wis., has 35 Brush lights, 17 on towers, the rest at the intersections of streets. Four towers have two lights each, and three have three lights each. The lights are burned on the Philadelphia schedule and cost $72 each per year. Contract five years.

FORT WAYNE, Ind., pays $150 per annum, with a discount of 10 per cent. each for 138 Jenney lights placed on towers in the outskirts and at intersections of the streets in central part of the city. The lights burn whenever needed and light seven square miles. Contract three years; wires overhead.

GALESBURG, Ill., has 96 Thomson-Houston lights placed on poles and at street intersections, which light $2\frac{1}{2}$ square miles. Thirty-one of the lights burn all night and 65 burn until midnight on the Philadelphia schedule. The all night lights cost $117 each per year; the short circuit lights $69 each; the contract is annual.

GALVESTON, Tex., 48 Brush lights placed on poles with overhead wires and lighted according to the Philadelphia schedule; cost 75 cents each per night, or $150 per year. Contract five years. There are also 199 gas lamps.

GLOUCESTER, Mass., has 15 Thomson-Houston lights on poles, in addition to gas and oil. The lights burn until midnight and cost $96 each per year. Contract annual.

THE CONTRACT, OR RENTAL SYSTEM.

HARRISBURG, Pa., has 150 Excelsior lights placed on poles which burn all night and every night. The contract, which is annual, fixes the price at $13,980, or $93.20 per light. Wires overhead.

HARTFORD, Conn., lights its central portion with 120 Thomson-Houston pole lights under an annual contract. The present price is five cents per lamp per hour. The lights burn all and every night, making the cost $198 per lamp per year. Wires overhead. Gas is used in the suburbs.

HAVERHILL, Mass., has 43 Thomson-Houston lights in the central portion, and gas, gasoline and kerosene on the outskirts. The electric lights are placed on poles, and burn until 1 o'clock. They cost 47 cents per light per night, or $171 per year. The contract is annual. Wires overhead.

HOBOKEN, N. J., has 100 Thomson-Houston pole lights within six square miles. The lights burn all night, and cost forty cents each per night. The five-year contract expires in 1893. Gas is also used.

HOLYOKE, Mass., the principal thoroughfares are lighted all night and the outskirts until midnight. Seventy-two Schuyler lamps are used in three square miles of territory. The all night lights cost 50 cents each per night and 37½ cents per light for midnight lights. The lamps are on iron poles; wires overhead; contract runs three years.

HORNELLSVILLE, N. Y., employs 64 American lights, placed at the intersections of streets, burning all night for $100 each per year. Contract three years.

HOUSTON, Tex., has just made a five years' contract for 100 Fort Wayne Jenney lights running all night for 41 cents per night per lamp, or $150 per year. Intersection cross arms are used, with overhead wires.

JACKSON, Mich., pays $16,000 per year for 180 Thomson-Houston lights, lighting three square miles on poles, with overhead wires, every night and all night. This is at the rate of $88.89 per lamp. The contract expired in March.

JACKSONVILLE, Ill., has 32 Thomson-Houston electric lights and 126 gas-lamps. The electric lights are placed on towers and mast-arms, and are of 2,000 and 1,200 candle-power. Four square miles are lighted 17 nights each month until midnight, and cost $100 each per year. Contract, two years. Gas costs $17.50 per lamp.

JANESVILLE, Wis., in addition to gas, burns eight Thomson-Houston lights until 11 o'clock and pays $100 each per year. The lights are placed on poles. Contract annual.

JERSEY CITY, N. J., has seventy pole lights, which are scattered in conjunction with gas. The lights burn all night, and cost forty cents each per night. The annual contract expires December 1, 1888.

JOLIET, Ill., lights four square miles with 100 American lights on poles, burning eight hours, 19 nights per month, paying 23 cents per night per lamp. Contract, five years; wires overhead.

KALAMAZOO, Mich., lights six square miles with 101 Thomson-Houston lights, placed at street intersections. Twenty-six burn all and every night, at a cost of 60 cents each, or $219 per year, and 75 burn until midnight for 40 cents each per night, or $146 per year. The contract runs three years.

KANSAS CITY uses 67 Thomson-Houston lights in dangerous places only. They are swung on mast-arms, with overhead wires, and burn all and every night, under an agreement that the price shall not be more than 55 cents each per night, or $200 per year.

KEENE, N. H., hangs 27 Thomson-Houston lights on ropes between poles or trees. The lights burn until 11.30 and cost $100 each per year. Contract three years. Gas and gasoline are also used.

KEOKUK, Iowa, uses 30 American lamps placed at intersections, burning all night for $140 each per year. The contract is annual. Gas is also used.

LA CROSSE, Wis., is lighted by sixty-two Brush lights, fifty of which are on poles and the others on four towers. The moon schedule is followed, and the cost is $861.12 per month. The five-year contract expires January 1, 1891.

LAFAYETTE, Ind., has 207 Brush lights, covering an area of five square miles, for which it pays 2 6-10 cents per hour per lamp, or last year $50.60 each. The lights are 600 feet apart, hanging in the center of the streets, with overhead wires, and burned on the Philadelphia schedule. Contract, three years, with privilege of five.

LANCASTER, Pa., pays 35 cents each per night, or $127 per year, for 138 United States all-night lights, which light two square miles. The remainder of the city is lighted by gas and gasoline. Pole system; overhead wires; annual contract.

LAWRENCE, Mass., uses 559 Edison incandescent and nineteen Brush arc lights, all placed upon poles. The lights burn all night, and the entire cost per year is $9,534.40. The five-year contract expires in 1890.

LIBERTY, Mo., uses 160 Heisler incandescent lights, placed on poles, and lighting one square mile. The lights cost $18 each per year, burning until midnight. There is no contract. No other light is used.

LIMA, Ohio, has just made a one-year contract for 58 Thomson-Houston lights, placed at street intersections, with one central tower and six lights, for $112 each per year. One-half of the lights burn all night on all but moonlights; the other half until 1 o'clock A. M. Extra lights may be ordered up at a cost of $100 per year each.

LOCKPORT, N. Y., has gas, arc and incandescent lights. There are 28 Brush arcs and 139 Edison incandescents placed on poles and swung over the streets. Three square miles are lighted, under a three-years' contract. The lights burn 2,000 hours during the year. The arcs cost $135 each, the incandescents $19 each, and gas $16.50.

LOGANSPORT, Ind., 62 Jenney lights burning on the Philadelphia schedule, swung on poles, with overhead wires, light four square miles. Price $115 per lamp per year; contract, annual; good satisfaction.

LONDON, Ont., has 62 Ball lights on two of the principal streets and at railway crossings and bridges. The lights are placed on poles 35 to 40 feet high, and burn all night and every night. Price 28 cents per light per night, or $102.20 per year for 27 lights; and 35 cents per night, or $127.75 per year, for 35 lights; wires overhead; contract, optional with the city. There are 350 gas-lamps.

LOWELL, Mass., has 125 Thomson-Houston all-night lights, for which it pays 55 cents each per night, or $200 per year. Poles are used, with overhead wires; contract runs three years.

LYNN, Mass., 100 Thomson-Houston pole-lights burn until 1 o'clock every morning, lighting two square miles. Price 47½ cents per light per night; contract, yearly; wires overhead; no other light is used.

MACON, Ga., lights 1¼ square miles with 34 Brush lights, placed on towers and intersection arms. The lights burn on the moon schedule, and cost $147 per year each. Contract, five years.

MANCHESTER, N. H., employs 40 United States lights in addition to gas. The lights are placed on mast-arms, and burn from 6 to 12 o'clock, and cost $132 each per year. There is no contract; wires overhead. Two square miles are lighted with electric lights; the remainder of the city with gas.

MANSFIELD, Ohio, has 74 Western lamps swung at street intersections, which burn until midnight. The cost is $75 each per year, under a five-years' contract.

MATTEAWAN, N. Y., uses 175 incandescent lights of the Heisler system, which burn all and every night, for $20 each per year. Contract, annual; wires overhead; pole system.

MASSILLON, Ohio, has 80 Schuyler lamps, which burn all night at street intersections, for $70 per lamp per year. Contract, annual; wires overhead.

MEMPHIS, Tenn., uses 75 Brush and Thomson-Houston pole lights, burning all night, at a cost of $180 each per year; wires overhead; contract yearly. The last legislature authorized the city to purchase a plant.

MILWAUKEE has just contracted for 20 lights at $150 per year.

MOBILE, Ala., has 106 Thomson-Houston lights, placed at alternate intersections in the suburbs, and at each intersection in the central portion of the city. There are also seven towers of four lights each. Four square miles are lighted for $14,500 per year, or $76.50 per lamp. On moonlight nights the lamps are not lighted. Wires overhead; the contract runs for five years.

MONTGOMERY, Ala., has just made a five-year contract for 100 Brush lights, placed at the intersection of streets, for 42½ cents per night per light, or $155 per year. The lamps burn all night and every night. The previous contract was for 60 cents per night; wires overhead.

MONTICELLO, Minn., uses 150 thirty-candle power Heisler incandescent lights, which are placed on street corners, and burn until midnight. They cost, under an annual contract, $24 each per year. No other light is used.

MONTREAL has 132 Thomson-Houston lights, burning all night and every night, lighting six of the principal streets three-fourths of a mile in extent. The cost is 60 cents per night per lamp, or $219 per year. Contract runs five years; pole system; wires overhead, and, in all cases, above other wires.

NASHVILLE, Tenn., the old contract for 30 Brush lights, at 70 cents each per night, has been renewed for 45 lights, at 60 cents each per night, or $219 per year. The lights are placed on an arch over the intersection of streets, and burn all night and every night; wires overhead.

NEWARK, N. J., uses, in addition to 3,562 gas-lamps, 184 United States lights, which burn all and every night, and cost 50 cents each, or $182.50 a year. Pole system; overhead wires; contract, three years.

NEW BEDFORD, Mass., uses 50 Thomson-Houston lights, placed on poles and burning all night, except moonlight nights, for 50 cents per light per night, or $182.50 per year. Contract, annual; wires overhead.

NEWTON, Mass., has 73 Thomson-Houston lights, 711 gas-lamps and 505 kerosene lamps. The electric lights are placed on mast-arms at the intersections of about ten lineal miles of the principal streets. They burn twenty nights per month, and cost 50 cents each per night. Annual contract.

NEWBURGH, N. Y., lights its suburban district with 83 Thomson-Houston lights, placed on poles, with overhead wires, and burning all night every night, for $120 each per year; contract, annual. Gas-lamps cost $32 each per year.

NEW BRITAIN, Conn., places 66 Schuyler lights on intersection mast-arms, and light one square mile. The lights burn until midnight; cost, $100 each per year. There is no contract. The city also uses naphtha.

NEW HAVEN, Conn., lights four square miles with 118 Thomson-Houston lights, placed on poles and mast-arms, with overhead wires. The lights burn all night and every night, and cost 50 cents each per night, or $182.50 per year; contract, yearly.

NEW ORLEANS has 30 Jenney lights in the public markets, 18 in the public squares and 771 for street lighting; a total of 819. Thirty square miles are lighted every night by poles and towers, for $125 per lamp per annum. Contract, five years; wires overhead.

NEW YORK has 24,719 gas lamps, 120 naphtha lamps and 831 electric lamps of the United States, Brush, Waterhouse and Thomson-Houston systems. The electric lamps light 34 lineal miles from dark to daylight, each and every night. They are suspended on ordinary poles 20 feet high, and cost 24 cents, 25 cents, 39 cents, 40 cents, 50 cents and 60 cents per lamp per night, according to the several contracts. The wires at present are strung overhead, but an underground conduit or subway is being constructed. The contract expires annually. New bids recently opened are as follows: Brush Electric Light Company, 441 lamps at 35 cents per night; United States, 411 lamps at 35 cents; Harlem, 36 lamps at 26 cents, 186 lamps at 28 cents, 156 lamps at 29 cents, 55 lamps at 35 cents, 19 lamps at 50 cents, and 7 lamps at 60 cents; Mount Morris, 22 lamps at 17½ cents, 79 lamps at 23 cents, 31 lamps at 28 cents, 52 lamps at 29 cents each, 129 lamps at 32 cents each, and 23 lamps at 40 cents each; East River, 493 lamps at 35 cents; Ball, 90 lamps at 27½ cents, on Broadway and Sixth avenue, from Twenty-third to Fifty-ninth streets; North New York, 238 lamps at 35 cents, 153 lamps at 29 cents each per night.

NORTHAMPTON, Mass., has 80 Thomson-Houston lights, of 1,500 candle-power each, which are placed on poles and mast-arms. The lights burn until midnight 25 nights per month, at a cost of $75 each per year. Contract, annual.

NORFOLK, Va., pays $80 each per year for 155 Brush-Swan lights, burning all night and every night. Pole system; wires overhead; contract annual. Four square miles are lighted.

NORWALK, Ohio, uses 82 Western intersection lights under a five-year contract, which cost $70 per lamp per year. The lights average six hours per night. Wires overhead.

OGDEN, Utah, has eighteen lights placed at street intersections, 675 feet apart, in the business sections, which cost $133 each per year.

OMAHA pays $180 per year for four lights at railroad crossings, and $206 each per year for eight lights burning all night on her viaduct.

ORANGE, N. J., has 426 gas lamps and fifteen electric lamps, which light one and one-half square miles at the street intersections. The lights burn all night, and the cost entire is $11,000 per year. The two-year contract expires December 1, 1888.

OSWEGO, N. Y., has 142 Remington lights which burn on poles at street intersections all and every night for 34 3-5 cents each per night, or $125 per year. The contract runs three years; wires overhead. Two and one-half square miles are lighted.

OTTAWA, Canada, pays $80 each per year for the first 70 of its 217 Thomson-Houston lights and $100 per year for each of the others, burned on the moon schedule and hung on poles 30 feet high. Contract three years; wires overhead. About three square miles are lighted. Poles are 600 feet apart; water power is used.

OTTAWA, Kan., has 40 Sperry arc lights and 200 Heisler incandescent lights placed on poles with arms. The Heisler lights cost $10 each per year, burned until midnight, and $16, burned all night. The contract runs three years.

PATERSON, N. J., uses 70 Arnoux lamps on swinging arms, which burn on the Philadelphia schedule. The cost is $90 per lamp the first year, $100 the second year, and $105 the third year; wires overhead.

PEORIA, Ill., with 225 lamps of the Jenney system, lights fifteen square miles, using towers and intersections. The cost is $145 per lamp per year, burned all night and on all but moonlight nights. Overhead wires are used. The contract is for five years.

PETERSBURG, Va., fifty-two miles of streets and alleys are lighted with gas and electricity. The Thomson-Houston system is used, furnishing 82 lights for $96 each per annum. Pole and intersection. Wires overhead. Contract expires in 1892. There are 297 gas lights.

PHILADELPHIA the past year has been lighted by 561 lights of the Brush, Thomson-Houston, Fort Wayne Jenney and United States systems. The lights are scattered over the city on poles owned by the city. Seven and a half miles of wire are underground in cables owned by the city, the balance overhead. The lights burn all night and every night, and the average cost has been 55 cents per light per night, or $200 per year. This was unsatisfactory, and bids were advertised for under a new contract, and ranged from 35 cents to 50 cents per light per night. A proposition for the city to own its own plant is now under consideration.

PITTSBURG employs 22 electric lights on her wharf for which she pays $185 per year.

PORTLAND, Maine, lights 3½ square miles with 168 arc lamps of 1,200 candle power each, and 250 incandescent lamps of 32 candle power each, for $26,000 per year. Extra arc lights cost $140 each per year and incandescent lights $18 each per year. The arcs are Thomson-Houston and the incandescents Edison. The lamps are burned from dusk to daylight each and every night. The pole system is principally used with overhead wires. The contract is for one year. No other light is used.

PORTLAND, Oregon, has 555 incandescent and 27 arc lights of the United States system. The incandescent lamps are placed on old gas posts and the arcs are hung by ropes over the intersections of streets. The lights burn all night and every night and cost $24.10 per year for the incandescent and $170 each per year for the arcs. The contract runs two years.

PORTSMOUTH, N. H., uses fifty Thomson-Houston pole lights, which burn until midnight, and cost $100 each per year. Gas and naptha are also used. The three-year contract expires in November, 1889.

POTTSVILLE, Pa., has 58 Schuyler lamps, 35 gas lamps and 20 oil lamps. The electric lights are on poles, burn all night and cost $105 each per year. Contract annual.

POUGHKEEPSIE, N. Y., lights the entire city, 2¾ square miles, all night and every night, with 187 Thompson-Houston lights, for which it pays $123 each per year. The lamps are suspended at intersections; overhead wires; contract annual.

PROVIDENCE, R. I., pays 50 cents each per night for 295 Thomson-Houston and United States lights burning all and every night. The pole system has been used, but in the new contract recently made the mast-arm is introduced. Wires overhead.

QUINCY, Ill., has 158 intersection Thomson-Houston lights, covering 3½ square miles, not burning on moonlight nights. Price $120 per lamp per year; contract for five years; wires overhead; general satisfaction.

RACINE, Wis., has 100 Brush and Thomson-Houston lights, part of which burn all night and the others until midnight.

For the former the annual cost is $70 per lamp, and $50 per lamp for the shorter circuits. Poles are used with overhead wires; contract annually.

READING, Pa., lights a portion with 107 Arnoux lights, paying 45 cents per lamp per night, moonlight nights excluded. Intersection system; overhead wires; annual contract. Intervening streets are lighted by gas. There is no competition.

RICHMOND, Va., has 118 Schuyler lamps placed on poles, burning all night for 40 cents each per night, or $146 per year. Wires overhead; contract for five years.

ROCHESTER, N. Y., has 617 Brush and United States arc lights and 700 twenty candle power Edison incandescent lights, burning all night and every night. Under the contract, which is for five years, the price for Brush lights is 30 cents per light per night for the first two years, 28 cents for the second two years and 27 cents for the fifth year. The price for the United States lights is $28\frac{1}{2}$ cents per light per night for five years. The incandescent lights cost 4 cents each per night the first year, $4\frac{1}{2}$ cents the second, $4\frac{3}{4}$ cents the third, $5\frac{3}{4}$ cents the fourth, and 6 cents the fifth. The pole system with overhead wires is used. There are also 1,750 gas lamps.

ROCK ISLAND, Ill., uses 11 towers with two Brush lamps on each, paying for each lamp $30 per month, or $360 per year. This was under a five years' contract. The service was not satisfactory, for the reason that but two lights were placed on each tower, and under a new three years' contract 32 mast-arm lamps have been added at the rate of $75 per lamp per year. The lights burn on the Philadelphia schedule.

ROME, N. Y., is lighted by 100 intersection Remington lights, which burn all night, for 45 cents each, or $164 per year. Three square miles are lighted, under a three-year contract, which expires next October.

ST. JOSEPH, Mo.—Proposals have been made for $180 per light per year.

SACRAMENTO, Cal., uses 36 Thomson-Houston intersection lights, which burn all night, except on moonlight nights, at a cost of $252 each per year. Contract runs two years; overhead wires. There are 193 gas-lights.

SAGINAW, Mich., lights three square miles with 66 Jenney lights, using 12 towers, 125 feet high, of four lights each, and

18 intersection mast-arms. The lamps burn until 3 o'clock A. M., and cost $125 per year for each tower light and $100 per year for each low light. Contract annual; wires overhead.

SALEM, Mass., has 143 Thomson-Houston lights, suspended on cables at street intersections, burning every night and all night. The price is $47 each for the first 100 lights and $45 for all over 100. Contract, two years; wires overhead.

SAN ANTONIO, Texas, a Brush plant has just been established. The city uses 60 arc lights, burning all night 24 nights in the month, at a cost of 60 cents per night per light. The system is pole and intersections; wires overhead; contract runs two years. At present three square miles are lighted. Gas costs $24 per lamp.

SANDUSKY, Ohio, five miles square, uses 125 intersection Western and Brush lights, burned on the Philadelphia schedule, for which it pays $82.50 each per year. Contract, annual; wires overhead. The plant was put in last November.

SAN FRANCISCO, Cal., lights a portion of the outlying districts with Brush lights. Twenty-one lamps, of 16,000 candle-power each, are placed singly on masts 150 feet high, and 17 lamps, of 2,000 candle-power, are placed on poles 20 to 40 feet high. Lamps burn all night, except on the night preceding and the night following full moon. The 16,000 candle-power lamps cost $5.28 per night, and the 2,000 candle-power 66 cents per night. Contract, annual; wires overhead.

SAVANNAH, Ga., lights four square miles with 100 Brush lamps, placed on towers and at intersections. The lamps burn all night and every night, at a cost of 70 cents each per night, or $255 per year. Wires overhead; contract, three years.

SCHENECTADY, N. Y., lights eleven miles of streets with 95 lights of the Remington system. The lights burn all night, are placed at intersections, with overhead wires, and cost 47 cents per lamp per night, or $171.55 per year. Contract, three years.

SCRANTON, Pa., has 200 Brush lights, swung on poles and at the intersection of streets, which burn every night and all night, under a three-years' contract, at 20 cents per light per night, or $73 per year. The wires are strung overhead. Four square miles are lighted. Gasoline is used in the outskirts, at $22.50 per light per year.

SEDALIA, Mo., in addition to gas, gasoline and coal oil, uses eighteen intersection lights, twelve of which burn all night for $145 each per year, and five burn until midnight for $120 each per year. The three-year contract expires July 10, 1889.

SELMA, Ala., has fifty Thomson-Houston lights, which swing at street intersections and light three square miles. The lights burn on the moon schedule and cost $120 each per year. The contract is annual.

SOMERVILLE, Mass., uses 70 American lights, on poles 20 to 30 feet high, until 1 o'clock every morning. Price, 37 cents per night per lamp, or $135 per year; contract, annual; wires overhead. Gas and oil is also used.

SOUTH BEND, Ind., has 20 United States lamps, placed on poles and scattered, with wires overhead. The lights burn until midnight every night in the year, for $100 each. The contract is annual.

SPRINGFIELD, Mass., 50 Thomson-Houston pole lights, burning all night, every night; cost 60 cents each per night, or $219 per year; wires overhead; contract, annual. Some private light wires are underground.

SPRINGFIELD, Ohio, has 54 intersection Thomson-Houston lights, which cost $130 each per year. The lights burn all night and light about four square miles. Wires overhead; contract, yearly.

STOCKTON, Cal., has just put in 100 Jenney lights, which light two square miles on the Philadelphia schedule. The lights are placed on eleven towers and 57 mast-arms. They cost $165 each per year. Two years' contract.

STILLWATER, Minn., lights 2½ square miles with 204 United States incandescent lamps, of 25 candle-power each, at a cost of $23 per light per annum; the lights are placed on gas-posts owned by the city, where such posts exist, and the moon schedule is followed up to 4 o'clock A. M. Wires overhead; contract for five years.

SYRACUSE, N.Y., has 280 Thomson-Houston lights, which burn all night, for $144 each per year. Pole system; wires overhead; contract, three years.

TAUNTON, Mass., has 20 Arnoux lights, which burn until midnight, and cost 50 cents each per night, or $182.50 per year. Overhead wires are used; lights at intersections of streets; contract, from year to year.

TERRE HAUTE, Ind., lights four square miles with 221 Thomson-Houston lights, placed on mast-arms at alternate street intersections. The lights burn 2,500 hours during the year, and cost $88.33 each. Contract, three years; wires overhead.

TOLEDO, Ohio, has two contracts. The first was made with the Brush Company at 55 cents per light per night, or $165 per year, when there was no competition; the second, with the Western Company, when competition was strong, is for $40 per year per light. Fifty intersection lights are used, burning all night, 300 nights in the year. Wires overhead; three years' contract; runs another year. The central part of the city only is lighted.

TORONTO, Canada, has 125 Excelsior lamps on a few of the principal streets. The lights cost 55 cents each per night, or $200 per year; are placed on poles, and burn 3,735 hours during the year. Contract runs five years. The wires are strung overhead, but the Council has been trying to secure legislation to compel the companies to place all wires underground.

TRENTON, N. J., on the principal streets are 76 Brush lights, placed on mast-arms, which burn all night and every night. Price, 50 cents per night per light, or $182.50 per year. Contracts for one and two years; wires overhead. Gas and naphtha is used in the outskirts.

TROY, N.Y., has 392 pole lights of the Brush and Thomson-Houston systems. They burn all night and every night, and cost 43½ cents each per night, or $158.77 per year. Wires overhead; contract, annual. There are 266 gas lamps which cost 10 cents each per night.

UNION CITY., Ind., has just made a five-year contract for Thomson-Houston lights, burning on the Philadelphia schedule, for $91.25 each per year. There are now 20 lights placed on poles at the intersections of streets, which light one and one-quarter square miles.

UTICA, N. Y., has recently entered into a contract with a New York company to light its entire territory with Jenney lights, on towers and intersections, for $42,000 per year, to the satisfaction of a committee of citizens appointed by the mayor. The cost of lighting the city in 1885 was $46,000, with gas and naphtha, and $48,000 in 1886 with gas, naphtha and electricity. There are 298 lights, and they burn all night. The contract is for three years.

VICKSBURG, Miss., has 40 Thomson-Houston pole lights, which burn all and every night, at a cost of $144 each per year. No other light is used. The contract is for two years.

WABASH, Ind., is lighted entire for one square mile by 127 Heisler incandescent lights placed at street and alley intersections. The lamps burn until midnight, and cost $2,300 per year. The five-year contract expires in 1893. No other light is used.

WALTHAM, Mass., has 32 Thomson-Houston lights, on poles, which burn until 1 o'clock. Oil and gas is also used. The electric lights cost 35 cents each per night, or $127 per year. Contract, annual.

WASHINGTON, D. C., but 87 public lamps are used. These are the United States, Thomson-Houston and Brush, and burn all night and every night, at a cost of 65 cents per night each, or $237.25 per year. The lamps are placed on poles, and the wires are underground in a terra cotta conduit. Contract annual.

WATERBURY, Conn., burns 90 arc lamps of 2,000 candle power each, and 48 lamps of 1,200 candle power each, of the Thomson-Houston system until 1 o'clock A. M. 26 days each month. The lights are on poles with overhead wires; light 3¼ square miles and cost 33½ cents each per night. The cost of full and divided arc lights is the same. Contract three years.

WATERTOWN, N. Y., has 102 Excelsior lights placed at the intersections of streets which light eight and one-half square miles, until 1 o'clock. The contract is for seven years, and the price $68 per lamp per year.

WICHITA, Kansas, uses 75 lights of the Thomson-Houston and Schuyler systems. They are placed at the intersections of streets, and burn until midnight, at a cost of $100 per light per year. There is no contract. The city also uses 120 gas-lamps and 300 gasoline lamps.

WILKESBARRE, Pa., has 32 Excelsior lights, which cost 39 6-10 cents each per night, or $144.54 per year. The lights burn all and every night, are placed on poles with overhead wires; contract for one year; gas lights cost $20; naphtha $18.

WILLIAMSPORT, Pa. The lowest bid received is $95 per light per year on a two-year contract, lights to burn every night and all night.

WINONA, Minn., uses 61 Van Depoele lights on towers and at street intersections, and lights twelve square miles. The lights burn all night on the Philadelphia schedule and cost $125 each per year. The contract runs five years.

WOBURN, Mass., has forty pole lights, which illuminate about two square miles. Eight burn all night, the remainder until midnight, and the entire cost is $3050 per year. Gasoline is also used. The two-year contract expires in May, 1889.

WOOSTER, Ohio, has 30 Schuyler lamps placed at intersections 400 feet apart, which cost $108 per year per light, burning all and every night. Gas and gasoline is used on the Philadelphia schedule. Contract two years.

WORCESTER, Mass., a few of the principal streets are lighted with 138 Thomson-Houston lamps, placed on bridges over intersections of streets, and using overhead wires. The contract is for three years and the cost 55 cents per night per lamp, burning each and every night, all night, or $200 per year.

YONKERS, N. Y., uses 50 Schuyler and Thomson-Houston lights placed on poles. The lights burn all night and cost $50 and $60 each per year. Contract three years. Gas is also used.

YOUNGSTOWN, Ohio, has 169 Thomson-Houston lights, within three and three-quarters square miles. The lights are on poles and mast-arms with overhead wires, burn until 2 o'clock A. M., according to the Philadelphia schedule. Price $64 per lamp per year; contract annual.

In every city but three the wires are strung overhead. The three exceptions are Philadelphia, Washington and Denver. In other cities there are underground wires in operation, as will be seen in the pages devoted to that subject, but so far as public lights are concerned, the three cities named are all. No element connected with electric lighting varies so much as the cost. There are almost as many different prices paid as there are cities enumerated.

Municipal Lighting.

THE entrance of municipalities into the field of electric lighting is comparatively a new element of competition which private companies have been called upon to meet. It is only within very recent years that the attention of corporations has been directed to the advantages of doing their own lighting on a principle similar to that on which they have been furnishing their own water. Investigations have been set on foot by many city councils, looking to the purchase and operation of a plant, and the question has risen naturally to others: If we can furnish ourselves with water cheaply, why cannot we also furnish light? Is one any more a governmental question than the other? Is it not good political economy as well as a sound business method? So general has been the consideration of this subject, that it will undoubtedly soon become, if it has not already, a leading topic in municipal ethics.

It can scarcely be gainsaid that, kept aloof from the intrigues of politicians and the influence of corrupt agencies, the possession and operation of an electric lighting plant for public purposes is a commendable investment for a city. The system should not be condemned because some of the barnacles that may attach themselves are undesirable.

The purchase of plants has thus far been largely confined to smaller places, which have required smaller plants and a lesser outlay. Some of the cities in this class have operated their plant long enough, however, for a demonstration of practicability and economy, and the information obtained is set forth in the following pages:

AT BAY CITY, MICH.

October 13, 1886, the street lighting committee of the Bay City, Mich., Council, presented a report of its investigations into the subject of buying a plant and doing its own lighting. So much of this report as is pertinent is herewith given:

"After a full, deliberate and impartial consideration of the subject, and with the additional aid derived from a complete investigation by other members of the council and city officials, and with the most earnest endeavors to act for the best interests of Bay City, to secure the best light for the least money, and to make no agreement unless fully guaranteed and warranted to indemnify the city against all losses, we respectfully report in favor of the plan that the city purchase, operate and maintain a plant of its own. As to the power of the city so to do, there is no question, as special provision therefor is made in the charter, and it is a function natural to and fully within the powers and attributes of a municipal corporation. Our reasons are substantially as follows:

"The amount of the capital supposed to be invested by the city in this enterprise, including the interest upon the sum invested and the cost of operating and maintaining the same, would effect a saving to the city as compared with the cheapest light offered on the basis of an annual rental, of about $5,000 per annum. We are of the opinion that the saving to be effected by the adoption of the plan we recommend, would, in the course of six years, pay the expenses of operating the plant, also the interest on the investment and the purchase money invested in the plant. Or, in other words, that the sum so expended would, at the expiration of that time, be equivalent to the rental * * * and the city would in addition become and remain the owner of the plant, besides having during that period furnished its own light. The city would thus be the gainer by the amount invested in the plant; whereas, on the rental system, the money paid out for the use or service rendered would be irrevocably gone without anything remaining over and above the light furnished.

"Having come to the conclusion that it is clearly for the interests of the city to own, operate and maintain its own plant, provided that it can be made sure without possibility of doubt or mistake, that a saving can be effected as above explained, your committee had under consideration * * * several bids * * * for furnishing the city a complete electrical plant, including all the items and apparatus requisite for the production of the best electric arc light possible, including building, boilers, steam engines, dynamos, etc. * * *

"The Jenney [the lowest and the successful] company warrant that the cost of running the plant of 120 lights, repairing and maintaining the same, shall not exceed $6,000 per annum. The actual estimated cost is only about $4,500 per annum. In support of these figures the committee found at one place they visited that the actual cost of running and maintaining a thirty-light plant was only $1,000 per annum; at another place forty-seven lights cost only $2,500 per annum; at still another, sixty-four lights cost annually $2,900. It should be borne in mind that these lights are operated by the city, and on all dark nights, rainy or cloudy weather, irrespective of the phases of the moon. All these plants are owned by the city, and the facts regarding them were obtained from the city officers. Such facts irresistibly forced the conclusion, that $4,500 per annum is a reasonable estimate of the annual expense of running 120 lights.

"The charge, that the efficiency and economy of the conduct and maintenance of the plant will be affected injuriously by political appointments, which will become the objects of partisan strife and contention, has little or no weight in this case, as the company making the proposition insists on the condition of naming or approving the superintendent in charge of the works.

"In conclusion we would state that, in every city we visited which was operating its own plant, we found everybody perfectly satisfied. It was the same story everywhere. Violent opposition and bitter fight at first—adoption of the system and purchase of a plant by the council—small expense for operating — great saving to the city — good lights running whenever wanted—everybody satisfied—would not be without it at any cost."

Bay City, now, has twenty-nine lights placed on towers, and ninety-one on mast arms at street intersections. These light five square miles until one o'clock A. M. The plant cost, complete, $30,280, has been running a little over one year, and is operated through a board of commissioners appointed by the council. The statement of the secretary and treasurer of this commission shows the average cost per light to be $39.60 per year. The system is the Indianapolis Jenney, and the wires are strung overhead.

Mayor Hamilton W. Wright, of Bay City, says:

"* * * Our plant is entirely satisfactory. We are running 120 lights, at a maximum cost per annum of $39.60 per light. We had 100 mast arm lights, and five towers 125 feet high with four lights each. Recently we purchased a high tower * * * and transferred nine mast arm lights to it, discontinuing the same number of mast arm lights, the total number remaining the same. We had been renting eighty-seven lights at $100 each per annum, making a total of $8,700 against $4,752 at present. Allowing interest at five per cent. on investment of $30,280, which equals $1,514, and two per cent. for wear and tear, which equals $605.60, we have a total cost of $6,871.60 for running 120 lights, as against $8,700 for renting eighty-seven lights. We would not abandon our present system on any consideration. Even those who most bitterly opposed it are now entirely reconciled and satisfied."

LEWISTON, MAINE.

Lewiston, Me., owns a 100-arc light plant of the American system. The lights are placed on street intersections, with overhead wires, and burn all night and every night. Further information is given in the following communication from D. J. McGILLICUDDY, Mayor of Lewiston:

"* * * The city of Lewiston has an electric lighting plant, owned and operated by the city. The plant is now in process of construction, and will be completed, ready to light the city, in about three weeks.

"We have heretofore lighted our city by electric lights partly, and partly by naphtha and gas. About four years ago we began lighting portions of the city by electricity. We let the contract to a private company, who put in a plant here, and it cost us from 55 cents to 65 cents per light per night, arc lights running till 12 o'clock. With our own plant, as per our estimate, it will cost us only 14 cents per night per light, lamps running all night. We shall run our plant by water power, which the city also owns. Our plant will be 100 arc lights, and all complete, ready for operation, will cost the city about $14,500; this includes $1,800 for a water wheel."

A month later Mayor McGILLICUDDY writes:

"Our electric lighting plant has been in operation now a little over a month. It gives the very greatest of satisfaction, and we would not think of such a thing as going back to the old system, even if they would give us the lights cheaper than we can get them ourselves. We find it a great advantage to have control of our own lights, and again we get our lighting done so much cheaper. Our lights cost us, running all night, less than 15 cents per light per night, reckoning interest on the cost of plant and all expenditures of every kind, including wear and tear of machinery, and we get far better light and better results in every way than when our city was lighted by a private company."

HANNIBAL, MO.

Mayor S. F. RODERICK, of Hannibal, Mo., writes:

"Our plant is owned and operated by the city. * * * We are a city of 15,000 people. Our present number of lights is 98—52 of these are for lighting the city and 46 are rented for commercial purposes. Of the city lights, 44 are on eleven towers and 8 on poles and mast-arms.

"The total cost of operating last year, every item of expense, was $6,282.92. Our income from rentals was $4,219.20, leaving a total expense for our city lighting of $2,063.72, for an all-night light, of unequalled brilliancy, and giving the greatest satisfaction to the people, as there is not a dark street in the city, and we claim the best-lighted city in the world.

"These figures and statements are *facts*, and should prove to any one the utility and advisability of each city owning and operating their own lighting plant. Then, if there are any changes to be made, lights to be added or lights to remove, there is no additional expense, only of wiring and cost of lamp. All profits accrue to the people, where they rightly belong. You always have the matter under immediate control, and can so conduct it as to be of the greatest good to the greatest number."

The system is the Fort Wayne Jenney.

PARIS, ILL.

J. M. BELL, Mayor of Paris, Ill., gives the following particulars of the subject for his city:

"We could not get an offer to light our city for less than $3,000 per year, such as we have now. The city owns the water-works, as well as the electric light plant, and both plants are built together, and we use the same boilers and building for both. Consequently it is run cheaper than to have them separate. One engineer and one fireman run both, and one superintendent runs all; so, you see, that we cut considerable expense on that score. We have the Fort Wayne Jenney light. We have 52 lamps, including four towers, run by two 38-lamp dynamos, with one Bass engine of 70-horse power. We have ten miles of wire, in a city of 6,000 inhabitants, and have one of the best-lighted cities that I have seen anywhere. * * * The expense last year was a little over $2,300. I do not think that the expenses will exceed $2,000 this year. We had been paying from $3,500 to $3,800 for lighting the city by gas, and not one-third as good a light. We think we can run our light for about $1,800 per year. Our citizens are well pleased with our light, and it gives entire satisfaction. * * * We figure that the saving from what it now costs us, and what we originally paid, will in seven years pay for itself, interest and all."

D. D. HUSTON, a member of the Paris Council at the time the plant was purchased, writes:

"Our city owns a plant consisting of an 80 horse-power engine, boiler, two 30-light dynamos, with 37 drop-lamps hung at the intersections of streets in the thickly-settled or business part of the city, and with four towers, 125 feet high, with four lamps each. Our light has given general satisfaction, and by use of the towers the alleys and outskirts of the city are as well lighted as the central part of the city. Our plant is operated in connection with our water plant, it being owned by the city, enabling us to operate our plant somewhat cheaper than were it alone, as our pumps are always in operation, so our light can be turned on at any moment, not having to wait to fire a boiler. It also enables the city to operate the light at a less expense than an individual or company could, not owning machinery that is always in motion.

"Many reasons pro and con were advanced before our plant was purchased. First, that it would have a tendency to purify city politics, as these light companies are generally composed of sharp, shrewd men. Their stock is distributed where it will do the most good in making city contracts, sometimes Aldermen and even Mayors being interested. It was observed, also, the company took a special interest in city elections. Men who never seemed to care who was made Legislator, Congressman, Governor or President, would shell out their money, go into the wards and voting precincts and spend their time and money to elect a man Alderman who never had any credit or standing in the community he lived in. It was a common thing for Councilmen to burn free gas, sprinkle his streets and lawn with free water, or such that the city was paying for. There was a constant issue of this kind. Politics cut no figure. The question was: 'Are you for the light company?'

. "Go into any city where light is furnished by contract and get the sentiment of the tax-payers, and you will find they are not satisfied, but tolerate it, fearing it might be worse. It is not always economy we want, nor the expense that tax-payers complain of; it is paying for something you don't get. No board has yet been able to make a contract with a light company that gave satisfaction to the tax-payers. Most contracts are based on so many hours lighting, taken from the moon schedule. Frequently our darkest nights are when the moon should be bright; so, in this event, we can turn our light on, if for only a fraction of an hour. We have found it to be economy.

"Our light has been in operation here since Sept. 1st, 1885. While we have some five times more territory lighted than we had before, we find the expense only amounts to about two-thirds the cost as when done by contract, and it has been done to the satisfaction of the people, and with a great relief to the Board of Aldermen. Our people seem to feel an interest; and should anything go wrong and the light not be up to expectation, they do not feel that they are paying for something and getting nothing."

MADISON, IND.

MADISON, Ind., lights five square miles with 81 Indianapolis Jenney lights, swung on mast-arms at street intersections. The lamps burn all night and every night; wires overhead. The plant is owned and operated by the city.

JOHN A. ZUCK, City Clerk of Madison, writes:

"The plant, entire and complete, including the power, which was built here, cost the city about $18,500. It is owned and managed by the city at a cost of $4,500 per annum. The city has been lighted by electric light nearly three years, and we think we have the finest lighted city in the world. Previous to the adoption of the light, gas was used, and cost about $8,000 per annum. The electric light will cost for the present year very little more than half what the gas did, and is so far superior as a light that there is no comparison."

J. T. BRASHEAR, Mayor of Madison, writes:

"We have been operating our electric light plant twenty months. We have the Jenney system of Indianapolis. We have three dynamos of 30 lights each. We have 82 lamps, suspended about 30 feet high at the intersection of the streets. About fourteen miles of wire, divided into three circuits, each dynamo operating upon its own circuit. Our system for lighting the streets with gas and gasoline cost the city $8,000 per year. The last year, with electric light, cost $4,600, or $55 per lamp, making a saving to the city of $3,400 per year, and the light is so much better than the old system—in fact, there is no comparison. Our citizens are well pleased with the light. The City Council of Topeka, Kansas, visited our city the first of this month, and it was their judgment that we had the best-lighted city they had visited. They were so well pleased when they returned home, that they contracted for a 120-light plant. * * *"

TOPEKA, KANSAS.

The experience of TOPEKA in the investigation and purchase of its plant is thus described by City Clerk GEORGE TAUBER:

"Our city advertised for bids for the purchase of a 120-arc light plant. * * * After due examination by a committee of the Council to investigate the various kinds of systems, and having received from five to nine different proposals, the Council adopted the Jenney light, of Indianapolis, and awarded the contract to them (complete plant, except building), 120 lights, for $26,196, with the further guarantee that the cost per light per month for eight hours shall not exceed $4.50, and for all-night light, $6."

The report of the electric light committee appointed by the Topeka Council, upon which the plant was furnished, is as follows:

"We visited Kansas City, St. Louis, Terre Haute, Indianapolis, Columbus and Madison, Ind. We included in our examination the systems of the Thomson-Houston, Brush, Fort Wayne, American, and Jenney of Indianapolis. Each of these systems, * * * we saw in two or more of the cities named. Each of us made tests, reading and otherwise, of the different lights. * * * Several of the committee read ordinary newspaper print at a distance of 280 feet * * *.

"At Madison, Indiana, we found the city most beautifully lighted; that city having for several years paid over $8,000 per annum for gas and gasoline, and finally turned to electricity. This, notwithstanding the fact that the city owns one-fifth of the stock of the local gas company, owns and operates its own plant, and * * * their own station. The cost of the running expense of their plant as per statement now in the hands of our city clerk, for the first year * * * amounted to $4,500, a little less than $55 per light per year. They have had very little repairs and its citizens are perfectly satisfied with the light. * * *

We found that at Bay City, Michigan, the city is operating and owning its own plant; they have 122 arc lights; the running expenses for 122 lights was $5,125.26 per year, or $42 for each light per annum, the plant costing $30,000.

"The city of Aurora, Illinois, owns and operates its own plant (Thomson-Houston system). They have seventy-two lights, 2,000 candle power, the running expense being $4,200 per annum, or about $57.33 per lamp per annum.

"We find that the Schuyler system in use at Wichita as per statement of our city clerk, for 250 lamps of 2,000 candle

power, is about $59.40, or $4.95 per lamp per month. * * *
"Your committee find the operating of plants by cities is meeting with great and increasing favor; in fact, the sum saved over ordinary rental prices will, it is estimated, in the course of five or six years, cover the value of the plant and operating expenses; meanwhile it appears to us there can be no question that for the best interests of the city of Topeka, we would recommend that the city own and operate its own plant and light by electricity, because we are satisfied that from information, figures and facts gathered, the light can be produced at about one-third of the best rental price yet offered, * * *

The bids were all based on the city engineer's specifications, and were as follows:

Brush Co., with high speed engines	$24,925
Brush Co., with low speed engines	25,140
Guaranteed limit of cost of lighting per annum:	
For all night every night	10,080
By moonlight schedule	9,640

Indianapolis Jenney Co., high speed, two 12x12 Ball engines	$26,496
Low speed	27,430
Guaranteed limit of annual cost.	
All night lighting $8.00 per light per night	11,520
Moonlight schedule $6.00 for light per night	8,640

Thomson-Houston Co., with high speed	$26,950
With low speed	27,775
Guaranteed limit of cost of running:	
All night	7,000
Moonlight schedule	4,800

Western Electric Co., high speed	$23,210
This company estimates the annual cost exclusive of coal:	
All night	4,300
Moonlight schedule	3,800

Schuyler Co., with high speed	$24,663
With low speed	25,091
Cost of running:	
All night	7,700
Moonlight schedule	5,500

CHAMPAIGN, ILL.

H. L. NICHOLET, City Clerk of Champaign, Ill., writes:

"This city does not own the electric plant in use here, but has the option of purchasing within six months from the time the plant was in successful operation, if satisfactory. The plant (Western Electric) has not been in operation quite a month, and, so far, gives good satisfaction. We have at present forty 2,000 candle power arc lights, the rate for same being $80 per light per annum. The lamps are hung upon mast-arms. The price of plant, should the city purchase, is $4,700.50, with reduction from same of rental paid up to purchase of same. Should city purchase, the company agree to furnish the power, and run it for the sum of $1,800 per annum, which is $45 per year for each light."

Mayor L. S. WILCOX, of Champaign, writes:

"Our city is lighted by 40 arc lights of the Western Electric Company's, Chicago, system. Lights are suspended by mast-arms at intersection of streets. The cost to the city of the 40 lights, 45 light dynamo and 11 miles circuit, was about $6,000. The city hire the lights operated until about 12 o'clock at night every dark night for a total expense of $1,800 per year. Counting interest on the $6,000 at 6 per cent, the total cost to the city for their lighting is $2,160. The cost of these lights if rented from the company would be at least $3,200 per year."

HUNTINGTON, IND.

Concerning the plant owned by Huntington, Ind., S. F. DAY, Mayor, writes:

"Our city bought their electric light plant about two years ago from the Fort Wayne Jenney Co. We made no mistake when we decided to own and control the business. We use 50 lights; sixteen are on towers, and 34 swing on cross streets, for which we would have to pay a company from $135 to $150 per light. Our entire expense for running, per year, does not exceed $50 per light, or $2,500 yer year. Our report last year showed $2,134 for eleven months. So you can see a good saving. Besides we have our city lighted any time we choose without extra expense. We claim to have the best and cheapest lighted city in the world—at least, I have never heard it disputed. We were the first in this State to buy outright. Many since have imitated us. * * * A clear saving of from $85 to $100 on each light, and being independent of a company is a good thing to have in a family."

DECATUR, ILL.

M. F. KANAN, Mayor of Decatur, Ill., writes:

"We have an electric light plant, which is not now, but will, I presume, become the property of the city in the near future. Our plant is being operated in connection with our water works plant, which makes it much cheaper than if it was an independent affair, we having ample boiler capacity for both.

There is this advantage in a city owning its own plant—they can have light whenever they feel the necessity, as for example in cloudy weather, which may occur in the light of the moon, which might not be provided for in schedule hours of contract or rental system. I believe that our 50 lights can be operated at an annual expense of less than $2,500, as it is now being run."

AURORA, ILL.

Mayor GEORGE F. MEREDITH, of Aurora, Ill., writes:

"Our city owns and operates its own system of electric light for street purposes. We employ the Thomson-Houston system and our plant consists of seventy-five 2,000-candle power lamps, two dynamos of 50 lamp power each, an Armington & Sims high speed engine of 80 horse-power, a boiler, and other appliances to make the plant complete. The electrical apparatus and power for operating the same are located in the pumping station of our water works, the night engineer of which attends to both the pumping and electrical apparatus.

"The actual cost of operating our electric light plant as above for the year ending Dec. 1, 1887, was $4,200. When I tell you that for three years previous to the time when our city commenced to operate its own system of electric light for street purposes, it had paid the local Brush electric light company $8,500 per year for operating 26 lamps of 2,000 candle power each, no better argument can be made in favor of any city owning and operating its own system of electric light for street purposes."

MARTINSVILLE, IND.

R. H. TARLETON, mayor of Martinsville, Ind., is the single exception to the general run of the foregoing sentiments. He writes as follows:

"You ask in regard to the advantages of a city owning an electric light plant, rather than being owned by a company. I do not see much advantage in a city owning it. I would prefer a company furnishing us the light at reasonable rates and making its own contracts with the citizens. The lines often get out of fix; the dynamo very frequently needs repairs—it is pretty expensive, all things considered. I cannot recommend a city to purchase when they can get a company to take hold of it. The light is fine, and we receive about $1,000 per year for lights furnished business houses, which cuts down our expenses, but still I would prefer that the plant be owned by a company which would take off a great deal of trouble to the city. Our plant cost a little over $5,000.

MICHIGAN CITY, IND.

MICHIGAN CITY, Ind., owns a 50 light plant of the Indianapolis Jenney system. The lamps are suspended in the center of streets and burn until 1 o'clock A. M. H. A. SCHWAGER, City Clerk, writes:

"It is somewhat difficult to give the exact cost of operating our plant, as we use steam from the water works' boiler, and can only estimate the amount of extra fuel used. Our light was bitterly opposed at first, and a great deal of criticism from parties interested in gas and other lights was heard, but everybody is for the lights now. The plant cost $8,500, and the cost of maintenance runs from $2,000 to $2,500, which equals $40 to $50 per lamp per year."

PORTSMOUTH, OHIO.

PORTSMOUTH, O., owns a Thomson-Houston plant which cost $18,000. There are 93 lamps of 1,200 candle power each in use, each lamp costing $38 per year. The lights burn whenever necessary early and late, as occasion requires.

PAINESVILLE, OHIO.

For two years past the Western Electric Co. of Chicago has been operating a plant for lighting the streets of PAINESVILLE, O., at a cost to the village of $72 per light per year. Mayor S. K. Gray writes that the "light has proved very satisfactory and we now wish to extend the system, and it is believed that we can add 20 more lights and thoroughly light our village, and by buying the plant save a considerable part of what we now pay. Our Light Committee have the subject of purchase under consideration."

YPSILANTI, MICH.

Mayor CLARK CORNWELL, of YPSILANTI, Mich., writes as follows concerning the plant in operation there:

"Our city put in and are running an electric light plant of 61 arc lights. The electric plant and power cost about $12,-000. It costs $166 per month for running expenses. It is estimated that $2,500 per year will cover the whole running expenses of repairs, etc. We are about adding 17 more lamps, and may make it 30 during the present year, which will reduce the cost per light for running expenses.

"We have five towers, which certainly give better results than the same number of lamps (20) on poles, and mast-arms hanging the lamps at intersection of streets. I am satisfied the city can operate an electric light plant for less money than private parties will furnish light. We run lights 20 nights and all cloudy nights till 12:30 A. M."

GRAND LEDGE, MICH.

PRESIDENT J. D. SUMMERS, of Grand Ledge, Mich., writes:

"Our citizens called a meeting and petitioned the village council to investigate the subject of light. We did so. We saw the Brush, Thomson-Houston and the Jenney. I will give you my reasons why I think it advisable for cities to own and maintain their plant. In the first place, if it will pay a company a profit, it will pay a city a profit. Before our village bought the plant, we offered to let a five year's contract to any individual to light the streets for $1,000 per year. We could find no one who would take it for less than ten years, which we refused. We then bought the plant which cost us about $4,500, without the power, a 30-light dynamo and all the other apparatus. We have fifteen lights on the streets, and fifteen commercial lights. We pay $1,000 per year for running and maintaining the plant. We get a revenue of $900 per year for the commercial lights; so, when our plant is paid for, it only costs us about $100 per year to light our streets. We think it is the finest light made, the nearest like sunlight, the steadiest, and for out-door and stores, can't be beaten. It is the Jenney."

PHILADELPHIA.

In the annual report of the Electrical Bureau of Philadelphia for 1886, Chief WALKER said:

"In the last annual report from this department, a suggestion was offered as to the advisability of extending the electric light service to all parts of the city, in connection with which I would add that the lighting of our public thoroughfares by electricity has reached a point at which, I think, would justify the city in establishing its own plant for that purpose; the city owning the gas and water works would have the power necessary for the service, and would be at little additional expense for that necessary agent; the outlay for conductors, lamps, dynamos, and other machinery would be the greatest expense. Its first cost would, in a measure, be returned to the city by saving the amount, charged by private companies in excess of the actual cost of production, and also by the increased facilities offered for public lighting."

CHICAGO.

CHICAGO has purchased and is now operating a plant of 100 lights, with which it lights the Chicago river, wharves, bridges and slips. It was first proposed to hire the lighting done, and bids were asked. The lowest was $65 per light per year—this by the Western Electric Co. This was not satisfactory to Supt. John P. Barrett, of the Electrical Department, and the city purchased the plant and is now operating it at a cost of less than $50 per year for each light. The city now has under consideration a proposition to purchase a plant to light a portion of the business districts, and Supt. Barrett's estimate for the same is found in a report which he recently made to the council. The estimate is as follows:

"* * * I have quite thoroughly investigated the various systems submitted, relative to their fitness for the purpose, and have classified them under three general heads. Two of these are of the incandescent, and one of the arc variety of electric light, and are known as—
"1. The Central Station Incandescent System.
"2. The Municipal Incandescent System.
"3. The Arc Light System.
"In the Central Station Incandescent System the distribution is made from a central plant, in the most economical manner, by what is known as the three-wire system. By this method a large saving is made in the amount of copper wire used, and a more even distribution of current for either light, heat or power than by other methods. In order to thoroughly cover the needs contemplated in the order referred to, it will be necessary that the capacity of the plant should equal an average constant demand of 75,000 sixteen candle-power lamps in use, with connections, with at least the equivalent of 175,000 such lights. This form of wiring and system of distribution, when once installed, will thoroughly and completely adapt itself to these requirements. The system is an extremely flexible one, allowing expansion and contraction at any and all times, and connections may be made or discontinued on demand. Means are provided by which an accurate system of accounting between the corporation and consumer is as simple as with gas or water plants. The pressure, or as it is known in electricity, the potential of the incandescent system, is extremely low, but constant, and therefore the liability to create trouble is very small. The light is agreeable and steady, and far more healthful than that obtained from any form of illumination where the oxygen of the air is consumed. While possessing all the above advantages, the creation of such a plant, requiring as it would the tearing up of the principal and recently paved streets of the city, would render its cost a heavy burden.
"The following estimate is based upon a capacity of 75,000 lamps, averaging 840,000 lamp hours per day:

CENTRAL STATION INCANDESCENT.

Underground work, trenching, repairing, etc.	$1,000,000
Boilers, engines, steam-fitting, bolting, shafting, etc.	400,000
Electrical appliances and machinery	550,000
	$1,950,000

"Real estate, building, etc., for plant should be added to the above estimate, unless property already possessed by the city could be made available—perhaps at the water-works.

ESTIMATED DAILY EXPENSE OF MAINTENANCE.

Coal, 210 tons at $2.50	$ 525 00
Breakage, 1,050 lamps at 85 cents	892 00
Water, 10 cents per 1,000 gallons	35 00
Oil and waste	25 00
Removal of ashes	15 00
Meter department	50 00
Other official expenses	40 00
Twenty firemen, $2.00	40 00
Twenty engineers, $2.50	50 00
Twelve dynamo and regulator men, $1.50	18 00
Depreciation, repairs to dynamos and electrical apparatus	40 00
Depreciation underground plant	50 00
Depreciation steam	70 00
Incidentals not named above	20 00
Total estimated daily expenses	$1,870 50

"Total estimated expenses, $1,870.50, or about the equivalent of gas at 50 cents per 1,000 feet.

"THE MUNICIPAL SYSTEM.

"This, like the former, is an incandescent system, but peculiarly adapted for out-of-door purposes. Some of its advantages are that it cannot be blown out, and is perfectly steady in the most violent wind. The cost of cleaning is a minimum, and it is readily lighted or turned off from the station in an instant. The lights contemplated in the following estimates are of thirty-candle power, but if desired they may be lamps of various candle powers upon the same line. The electrical pressure in the municipal system is much higher than that in the central station incandescent system, but is below that in the ordinary arc light systems. It is not a proper system to introduce into dwellings or other places where possible contact would be made by accident. In underground conduits or pipes, however, and handled only by experts trained in its use, it is perfectly safe and readily controlled. The current, which in the arc system is concentrated at the lamps, is in this system capable of division into smaller lights, and these may be more frequently placed, thus having the advantage of more perfect distribution and more uniform illumination. There is abundant room for a station of this character in the quarters now occupied by the city river plant, should this system meet the approval of your honorable body, and an expense of say $8,000, would suffice for the necessary changes and construction. The estimate is based on an average lighting of seven and one-half hours per day.

THE MUNICIPAL INCANDESCENT SYSTEM.

Dynamos, electrical apparatus	$ 19,800 00
Engines, boilers, etc.	26,400 00
Underground conductors, delivered in Chicago	16,500 00
Gas-post extensions	19,800 00
Superintending and sundry expenses	8,500 00
Conduit, including trenching and repaving	442,698 00
	$533,698 00
Buildings for station as above	8,000 00
	$541,698 00

MUNICIPAL LIGHTING.

MAINTENANCE PER DAY.

Coal	$ 50 00
Oil, waste, etc.	3 00
Engineer and electrician	10 00
Assistants	9 00
Firemen	6 00
Linemen	5 00
Repairs	4 00
Lamp renewals	40 00
	$128 00

Cost of maintenance per year, $46,720.

"THE ARC LIGHT SYSTEM.

"This is the most simple of the systems, and is identical with that used by the city on the river, viaducts and bridges at Rush, Lake and Twelfth streets. The estimate is based upon the number of street crossings and middle of blocks within the district named. Of these there are in the northern division, 222; western division, 254; southern division, 231; a total of 707 crossings and middle of blocks, at which it is proposed to place a lamp of 2,000 estimated candle-power, with a capacity for all-night illumination, known as double lamps. Allowance is made for a limited excess above that number, the estimate being based upon 750-lamp capacity.

ESTIMATED COST.

Dynamos complete, with 750 double carbon lamps	$ 55,350 00
3,000 horse-power engines, boilers, pumps, heaters, etc., complete	30,000 00
750 posts, 16 feet, erected	11,250 00
75 miles underground cable, with iron pipe, at $1,500 per mile	112,000 00
75 miles trenching and repaving, and laying cable as above	218,630 00
Rebuilding present station	8,000 00
	$435,230 00

"The present gas-lighting schedule includes about 2,200 hours' lighting per annum, or an average of something over six hours per day. Assuming this as a basis of calculation, the following estimate of yearly maintenance is given:

Coal, 4,000 tons at $3.00	$12,000 00
Carbons	3,500 00
Oil, waste and incidentals	3,500 00
Chief engineer and electrician	1,500 00
Three assistants	3,000 00
Four firemen	2,400 00
Ten trimmers and linemen	6,000 00
Repairs	4,000 00
	$35,900 00

"Assuming that there are 3,273 public and private street lamps within the territory named, and that each of these sheds sixteen candle-power light, we have a total of 52,368 candle-power.

"With the incandescent system 3,273 lamps of thirty-candle power each would give 98,190 candle-power.

"Seven hundred and fifty arc lamps would shed, at 2,000 candle-power, a total light equivalent to that from 1,500,000 candles. Allowing for any possible candle-power claimed, there would be still a large balance in favor of electric lighting.

"It would seem from this that the illumination as between the incandescent and the arc is in favor of the arc, but the distribution is more uniform from the incandescent than from the arc.

"The street incandescent system contemplates the use of a low-pressure current, and a lamp of nearly double the candle-power of the ordinary sixteen candle-power lamp. These are to be placed on the present gas street-lamps, with extensions as shown in diagram as herewith presented. The arc lamp, on long circuits, because of its high tension character, is difficult of insulation restraint. The lamps require daily attendance, replacing carbons and maintaining the lamp machinery in order, without which these refuse to burn. The incandescent, on the other hand, is a light which requires a less pressure or potential of current than most of the arc lights, and insulation is comparatively easier on that account. There is no machinery to become clogged, no falling dust or sparks; only when a filament is worn out by use, or a globe is broken, is there need of a trimmer. The light is less powerful than that from the arc, but it is steady, soft and agreeable, and being enclosed from the weather, is not as liable to become grounded.

"A very material fact in this connection is the difference in the distribution of the light, which to me argues strongly in favor of the incandescent form of light for residence streets, where trees and foliage shade the light. With this system the same amount of light may be made to uniformly cover a large space, with a multiplicity of burners scattered over that space. The estimates above are based on such distribution at the lamp-posts now used for gas, thus giving nearly twice the candle-power at each of such posts. For points where powerful concentration of the light is required, as at the bridges, along the river, viaducts and parks, the 2,000 candle-power arc light is far preferable; but where the light can be divided so as to cover a greater extent of territory, and less illumination is required at some one point, the street incandescent system seems to me to be the more advantageous. The arc light should be used in the business portion of the city, and the incandescent in that portion of the city where foliage, shrubbery, etc., is located, and where more equal distribution of the light is required. * * *"

EASTON, PA.

Mayor Charles F. Chidsey, of Easton, Pa., writes as follows about the plant which his city owns:

"Our electric light is the Western. It has given us great satisfaction. It has run since August 2, 1886. We have a building for the plant, 40x80; two Ide engines of sixty horse power each; two boilers of seventy horse power each; three dynamos of twenty-five light power each; sixty-six arc lamps, suspended over intersection of streets, and twelve miles of wire. We have a paid superintendent, engineer, fireman and two line-men. Our plant costs the city to run it about $7,000 per year, or $105 per lamp per year. Our people vote it a success in every particular."

LYONS, IOWA.

Mayor J. C. Hopkins, of Lyons, Iowa, writes:

"Our system is a small one of the Ft. Wayne Jenney patent, put in and operated by the city. I have spent considerable time and have traveled a great deal with a view of investigating electric systems. I have seen almost all in operation practically, and I have had over a year's experience with the Jenney. I would not change for anything I have seen.

"I am equally firm in my opinion that every city should own and operate its own plant. Ours is costing us to-day an average of $30 per lamp per year. We have nearly as many lamps in the hands of private parties as we have street lamps, for which we charge $6.75 each per month, and, of course, that is deducted from the operating expenses in my estimate of net cost. Should we enlarge our plant, we could make it self-sustaining, but at present we are unable to furnish private lamps asked for. There are parties here to-day who stand ready with twenty-four hours' notice to take the plant off our hands at its full cost, providing a reasonable contract be made for lighting the city. However, should we enlarge our plant, I am satisfied it will soon be self-sustaining, and I think any city can accomplish the same result, providing exclusive privileges have not been voted to private corporations."

FAIRFIELD, IOWA.

T. F. Higby, city clerk, writes:

"Our light is run in connection with out water-works, using the same boiler for steam and the same engine. Both are owned and run by the city, under direction of a committee of the council. The system is the Brush, and there are but seven lights placed on a central tower. The lights burn until midnight. The estimated cost is $1,200."

LITTLE ROCK, ARK.

On and after July 1, 1888, Little Rock, Ark., will be lighted by a plant owned and operated by the city. There will be 90 Fort Wayne Jenney arc lights, 74 of them on mast arms at street intersections, and 16 on four towers. It is intended that four square miles will be lighted all night. The plant cost $27,000.

SHERMAN, TEXAS.

Sherman, Texas, has just purchased a plant of fifteen Van Depoele lights. The lights are placed on poles, and burn all night. The estimated cost is $80 per light per year. Gas is also used at present.

NORTHAMPTON, MASS.

A committee of the Northampton council, to whom the matter was referred in 1886, reported that the city could establish and maintain an electric light plant, furnishing arc lights of 2,000 candle power, running every night from sundown to midnight at a cost of $50 per year per light, on a liberal estimate, and the city clerk says that had the Council adopted the report the saving to date would have paid one-third the cost of the plant.

DETROIT.

In November, 1888, the board of aldermen of Detroit appointed a committee to investigate the practicability of the city purchasing and operating a plant. An exhaustive report has been prepared, a great deal of the information of which is contained in this book.

In summarizing the figures obtained from cities which are doing their own lighting, the report says:

"BAY CITY, Mich., runs its lights on the moonlight schedule until one o'clock A M., at a cost of $39.60 per light per year. The cost under a contract for the same service was $100.

LEWISTON, Me., with water power, runs all night and every night for $54.75 per light per year. The cost under the contract system was $200.75 and $237.25.

HANNIBAL, Mo., running all night and every night, costs $64.11 per light per year.

PARIS, Ill., could not get a contract for light for less than $3,000 per year. Gas the last year used cost $3,800. The cost of maintaining its plant, in connection with the water works, ranges from $1,800 to $2,000.

MADISON, Ind., paid $8,000 per annum for gas. It now maintains its electric lighting plant for $4,600.

The guarantee for the new plant TOPEKA, Kas., has purchased is, that the cost of maintaining an all-night light shall not exceed $96.00 and for a moonlight light $72. The Brush Company offered a guarantee that the all-night light should not exceed $84, and moonlight light $80.33.

The Western Electric Company agree to run the plant at CHAMPAIGN, Ill., if the city purchases, for $45.00 per light per annum.

The cost to HUNTINGTON, Ind., does not exceed $50.00 per light per year.

AURORA, Ill., paid the Brush Company $8,500 for twenty-six lights. Now it maintains seventy-five lights for $4,200.

The cost to MICHIGAN CITY for one o'clock lights is $40.00 each per year.

PORTSMOUTH, O., pays $38.00 for a divided arc light burning whenever necessary.

CHICAGO operates a plant for $50.00 per light per year, burning all night.

The cost to YPSILANTI is $32.33 for a midnight light.

The highest of the above is $64.00. If Detroit were to light her present system at that rate, the saving would reach $77,540 per year. If the cost should reach $100.00, which is more than one-third greater than any data shows, the saving would still be $55,040. For 1,000 lights, the standard adopted, the saving would, at $64, be $135,665 per year; at $100 it would be $99,665."

How to Buy a Plant.

THE electric light is unquestionably the most economical and effective means of illumination yet produced. Its superiority over gas and oil has met with full and public recognition. It secures freedom from impure and overheated air, from noxious and unhealthy vapors; it affords safety from fire when properly inaugurated, and yields a steady, brilliant light, which shows everything in its true color.

All cities and towns in the United States of any size will, sooner or later, be illuminated by electric lights. It is only a question of time, and of a very short time. This being the case, it is important that those intending to introduce this method of illumination should be posted, both as to the character of the light and the commercial value of the various systems. There are nearly a score of systems of electric lighting, good, bad and indifferent, which are being operated with more or less success, and it is of the first importance that the intending purchasers should examine them thoroughly, and become familiar with their workings before deciding upon which system they will adopt. A good way to lay the foundation for an investigation is to subscribe for and read one or more of the several electrical publications, all of which are filled with the knowledge which the searcher seeks.

There are several points in regard to electric lighting in general which it is well to remember. There are two general types of arc lamps; those working with a clutch, and those working with gear. The two schools have their advocates. Again, there are offered in the market lamps of different degrees of candle power. Some are 1,200 candle power, and some are 2,000 candle power, nominal. None of these lamps run at any such power. The average 2,000 candle power lamp gives about 800 or 900 candle power, and the half-arcs, or 1,200 candle power lamps, give about 500 or 600 candle power. Again, great stress will be laid upon the economy of power, some systems claiming to give a lamp for as low as each half horse power expended. The fact remains, however, that to give what is generally termed a 2,000 candle power lamp, at least nine-tenths of a horse power is required. Where the claim is made to furnish a lamp for little or no horse power, it should be understood that it is for a half-arc of nominal 1,200 candle power, or the parties are misrepresenting. It is not possible, by any reliable system of lighting, to give a continuous and satisfactory 2,000 candle power light by the expenditure of much less, if any less, than a full horse power of energy. On long circuits it will require more than a horse power per lamp, as a certain per cent. of current is lost within certain distances, as, for instance, a station which puts 1,000 lamps upon the streets, would be employing fully 1,200 horse power of energy.

Every company will lay stress upon its automatic regulation, and some will claim that they are the sole owners of the only method by which it can be accomplished. The truth of it is that there are several good systems by which it can be accomplished, and which are not owned by any one company.

The elements entering into a successful system of electric lighting differ in no respect from those of any other branch of business, and are briefly:

1. Economy in first cost.
2. Economy in operation.
3. Simplicity of construction.
4. Durability.

As to the first requirement, the time has come when electric light apparatus is placed upon the market at a cost no greater than that required for any other piece of machinery requiring equal skill and value to material of manufacture.

Economy in operation is attained when the maximum of light is produced with the minimum expenditure of power.

The advantage of simplicity of construction is that it gaurantees freedom from annoying interruptions.

The question of durability is a difficult one to solve, and one upon which nothing but the experience of others should have weight. The first consideration of course is that the apparatus should have proper care, such as is given to any other piece of good machinery. If the station is not in good hands, do not charge everything to the system. The problem before electrical engineers has been not simply to produce a good light, but also to produce an economical light, and these meritorious efforts should not be frustated by careless or ignorant handling of the apparatus.

As is now generally known, electric light is produced simply by expenditure of power; consequently, that system which will produce the largest quantity of light from a given amount of power, must be the most economical and prove the greatest success commercially. So it is that the first cost alone does not determine the economy of a system; the most economical system is one whose running expenses for power and attendance and depreciation are least.

Without regard to description, whether steam, water, gas or other power supplied, there should be at least one-fourth more than the station ordinarily uses. This will prove most economical and insure good results. Never should any attempt be made to run electric lights where there is lack of power; it is simply waste of time, useless expense, and causes general dissatisfaction.

In selecting engines for electric lighting, none but the very best should be adopted. The price may be somewhat higher than the common engines, but ample reward will be given by the superior results obtained. Since electric lighting has come in vogue, steam engineers have turned their attention to the want, and very fine engines have been brought in the market. There are two schools in this field, the advocates of high speed and of low pressure assuming that each is pre-eminently the best. Another and important item is, that steam should be kept as near as practicable at even pressure. In cases where the power is taken from a main shaft of a shop, it should be steady, since variation of speed will produce variation of current, and all will reflect in the light.

When the speed of a dynamo has been determined, it should be kept there as nearly as possible. It is in all cases advisable to use independent power to drive electric-light machinery.

The speed will be regular, and the light can be run while the machinery is at rest. Let it be remembered that the current in a dynamo corresponds exactly with the power which drives it. When the power is steady the current will be steady, and perfect success is secured.

To install a plant properly, so that all the lights burn with equal brilliancy and so that the loss of energy in the circuit or "line-wires" is reduced to the lowest practical limit, is a problem requiring, in addition to a thorough theoretical knowledge of the subject, a large and varied practical experience in the erection of plants. This is a most important matter to the purchaser, as, where there is competition, advantage might be taken of the general ignorance of the subject to cheapen the cost of construction by loose work or inferior material, which will soon make the expense of operating the plant greater than it should be. The purchaser should make as thorough an investigation as is possible before deciding on a system, and should bear in mind the fact that a good dynamo, a good lamp and a good steam plant, though all are necessary, by no means insure the satisfactory operation of a plant. The certainty of having it properly installed should receive as much consideration.

When the subject has been fully canvassed, complete specifications of just what is wanted should be made out and sent to all reputable electric-light companies, asking for bids on the same. There is a wide difference of opinion as to what constitute proper specifications. Electric engineers differ as widely as do those not so conversant with the technical side of the question. In all cases the purchasers should prepare or cause to be prepared their own specifications, and not leave the task to any particular companies.

For the information of contemplating purchasers some sample specifications are appended. They are divided into steam and electrical divisions, and these in turn are divided into the two systems, from which selections must be made. The plant under consideration is for 1,000 arc lights of nominal 2,000 candle power, or 4,000 incandescent lights of nominal 50 candle power, on the streets. Multiples of this can be made to suit the size of plant required.

· (Illustration.)

HIGH SPEED STEAM PLANT.

ENGINES.—Twelve engines of 125 horse power each, engines to be of high speed, automatic cut-off, with hammered wrought iron or cast steel shafts. Each engine to be provided with throttle valve, two band wheels turned true and balanced, oiling devices of the most approved type, consisting of side feed lubricator, sight feed oil cups with stands, needle valves and wipers, full set of wrenches, foundation rods, anchor plates, etc., to cover complete engine ready to receive steam and exhaust pipes. Each engine to have a heavy cast iron self containing sub base or coping stone. Said engines must develop the given horse power under an initial pressure of eighty pounds to the square inch.

BOILERS.—Twelve tubular boilers, each seventy-two inches in diameter, containing sixty-four four-inch tubes, sixteen feet long, with shell extension of not less than twelve inches. Each boiler to be provided with two man holes, one located on top of shell at rear end, the other in front of heads below the tubes, said man holes to be not less than ten inches by sixteen inches. The material for shell to be best quality homogeneous flange steel, sixty thousand tensile strength per square inch of section, and not less than three eighths of an inch in thickness. Material for heads to be of same material and of same tensile strength, and of not less than nine sixteenths of an inch in thickness.

NOZZLES.—Each boiler to be provided with two nozzles riveted to shell, and of proper size for steam pipe and safety valve connections, and of not less than six inches in diameter.

BRACING.—Heads to be thoroughly stayed to shells, by long overlapping braces. Each stay-rod to be made of best Swedes iron of not less than one inch in diameter, and attached with solid crowfoot and strap head properly riveted to heads and shell plates.

SEAMS.—Seams to be tight fitting, and rivet holes punched to line true one with the other, and no drifting allowed to force holes in line. Holes out of truth to be reamed with sharp edged cutting tool and rivet fitted to hole and driven straight and true. All joints to be planed or chipped and caulked. All longitudinal seams to be double riveted.

TESTING.—Each boiler to be tested to one hundred and fifty pounds per square inch hydrostatic pressure.

BOILER FIXTURES.—Each boiler to be furnished with a full line of fixtures complete, including full fire fronts, liners, grates and bearing bars, return draft arch, ash doors and frames, wall binding rods and back stays, cast iron wall brackets for supporting boiler, with wall plates and expansion rollers, all necessary gauges, steam and water stop valves, blow off check and safety valves, and firing tools. Each boiler to be set in separate arch and to be entirely independent of others.

PUMPS.—Two duplex steam lift and force pumps each capable of supplying twelve boilers of a capacity of one hundred and fifty horse power each.

INJECTORS.—Two injectors as auxilliaries to steam pumps for feeding boilers, each capable of delivering 7,500 gallons of water per hour at a steam pressure of eighty pounds.

HEATERS.—Two brass tube feed water heaters, each capable of supplying a boiler capacity of eight hundred horse power.

FURNACES.—Twelve furnaces (state whether with or without automatic feeders and stokers), placed under boilers in batteries of six each connecting with a brick chimney eight and one-half feet inside diameter and not less than one hundred and forty feet high. Said furnaces to be erected by the bidder, with the chimney.

GRATES—Grates to be either rocking or stationary (state which) and not to have less than fifty per cent. air space, with openings of not more than one-half inch.

STEAM PIPING.—From boilers to engines one main line of steam pipe, with straight-way stop valve, pipe to be of sufficient diameter to supply twelve engines each of 125 horse power, with branch pipes of sufficient diameter for each engine in each of which there is to be straight-way stop valves. In leading pipes from boilers to main line there is to be straight-way stop valve placed near connection of said pipes with boilers.

EXHAUST PIPES.—The exhaust pipes of each engine to be run to a main exhaust pipe leading to heaters and this main pipe to be of sufficient diameter to allow of free exhaust from eight hundred horse power. Straight-way valves shall be placed in each branch exhaust pipe leading to the main. From each heater there is to be an exhaust pipe leading to top of building of not less than sixteen inches diameter.

WATER PIPES.—Steam and water pipes between main steam lines and pump and injectors, and all water lines between water works connection to conform to tabulated sizes required for pump as heretofore specified.

PROTECTION.—All steam exhaust and water pipes to be thoroughly protected with sectional pipe covering.

DUTY.—The bidder to estimate on preparing all foundations and masonry and to deliver, erect and connect all engines, boilers, heaters, pumps, etc., and turn the steam plant complete over to the purchaser ready for duty.

ESTIMATES to be in detail, stating name and kind of each article and price, and guarantee amount of fuel per horse power delivered.

CORLISS STEAM PLANT.

ENGINES.—Three compound condensing Corliss engines, each capable of developing six hundred horse power, with a boiler pressure of ninety pounds. The material and workmanship to be first-class in every respect. Bidder to state diameter and stroke of cylinders, diameter and length of engine shafts, diameter and length of main bearings, diameter, face and weight of fly wheels. Each engine to be provided with a full set of graduating oilers, sight feed cylinder oiler and set of hand oilers with tray. The oilers to be so arranged that they may be refilled while the engine is in operation.

AIR PUMPS AND CONDENSERS.—Three air pumps and condensers of the spray or jet type, each to be connected to one of the above engines and to be of proper capacity for the engine. Bidder to state diameter and stroke of air pump, diameter of suction and overflow pipes, and state whether the air pump is driven by belt, connecting rod or independent steam cylinder.

BOILERS.—The boilers (state kind) to be of sufficient capacity to furnish the steam required to run the above engines up to 1,800 horse power, each boiler to be as follows: Sixty inches diameter, sixteen feet long, containing forty-four tubes, four inches in diameter; shell of boiler eleven thirty-seconds of an inch thick; heads one-half inch thick; dome thirty inches in diameter by forty-four inches high, double riveted to shell; shell of dome five-sixteenths inches thick; head of dome seven-sixteenths inches thick; all to be homogeneous steel, 60,000 pounds tensile strength. Longitudinal seams of boiler and dome to be double riveted. Two cast iron brackets, riveted to each side of boiler. Boiler to have one manhole on top of shell and one manhole in front head under the tubes. One full front with anchor bolts, five feet grate bars, bearing bars, liners, ash door and frame, back plate for arch, binder bars and rods. One set of boiler fixtures, including three inch pop safety valve, two inch blow off cock, one and one-half inch feed valve, one and one-half inch check valve, steam gauge, glass water gauge and gauge cocks. Suitable breeching of No. 12 iron to connect all the boilers to brick stack.

STEAM PUMPS.—Two steam pumps, each large enough to feed the boilers required to run two of the above engines.

PIPING.—All necessary pipes and valves inside of boiler and engine house to connect above engines, condensers, boilers and pumps. The steam pipe from all the boilers to be connected to one header. From this header one supply pipe to run to each engine. The exhaust pipes to be so arranged that the engines can be used non-condnsing when necessary.

SHAFTING, ETC.—(State whether one or two) line shaft of suitable diameter and length to drive forty dynamos, each of thirty arc light capacity. A suitable number of ball and socket pillow block bearings to support above shaft. Three main driving pulleys, one to be driven by each engine, these to be arranged with hollow sleeves and jaw clutches. Forty friction clutch pulleys, each of proper size to drive one dynamo, each clutch to be provided with a lever for throwing clutch in and out.

BELTING.—(State whether one or two) belt from each engine flywheel to the corresponding driving wheel on line shaft to be best short lap, oak-tanned double leather belting. Forty dynamo belts, one from each of the friction clutch pulleys to its corresponding dynamo, to be best light double leather dynamo belting.

MASONRY.—The engines, condensers and feed pumps to be provided with suitable foundations of brick or rubble stone, with cut stone cap stones. The shafting to be supported on piers built of brick with iron cap plates. The boilers to have substantial brick settings. All the inner surfaces of the walls exposed to the flame to be lined with first-class firebrick.

TESTS.—The bidder to guarantee the number of pounds of water evaporated per hour per indicated horse power required for each engine when developing six hundred horse power, and also to guarantee the evaporation of the boilers using (state kind of) coal.

ARC LIGHTING PLANT.

DYNAMOS.—Forty dynamos of thirty arc lights each, to be mounted on adjustable insulated bases with belt tightening apparatus. Each dynamo to be provided with approved lightning arresters, automatic regulator, to fully govern and control the current, and regulate the same in proportion to the number of lamps in operation from one lamp to the maximum capacity of machine, a test of twenty-four hours to be guaranteed. Each dynamo shall be provided with approved ampere meters, and the terminals shall be taken to suitable switchboards, so arranged as to enable the operator to manipulate the dynamos and circuits at will, with the least possible opening of circuit. Bidder to state lowest guaranteed horse power required in each arc and to drive each machine. Lamp magnets and dynamo armatures must be guaranteed against burning out from short circuiting of the current for two years. Dynamos shall admit of being coupled together to operate any multiple of their capacity.

LAMPS.—One thousand double carbon, standard 2,000 candle power arc lamps, each lamp to have automatic cut-out, and hand switch, with lifting and sustaining magnets of sufficient strength to enable the use of carbons one half inch in diameter if desired; said lamps shall be thoroughly insulated and protected from the weather by hoods or other devices; bidder to state the nature of such devices. Bidder must state the number of amperes of dynamo running under normal condition, also the voltage per lamp.

LINE WIRE.—(State number) miles of No. 4, 5 or 6 B. & S. Gauge copper line wire, water, weather, and fire proof insulation, said wire to be subject to test of twelve hours under water, after which it must show (state) resistance, and shall be tested from reel supplied for work. Said wire shall not be less than ninety-eight per cent. pure copper. Every joint to be soldered and wrapped with tape.

POLES.—Four thousand forty-five foot cedar poles, not less than six inches in diameter at the tip. Said poles shall be set six feet in the ground and covered with preservative, and the part above ground trimmed and painted with two coats of paint.

CROSS ARMS.—Six thousand four-pin cross arms, four and one half by five and one half inches by four feet in length, free from knots and cracks, and painted with two coats of paint.

INSULATORS.—Twenty-four thousand deep grove electric light insulators.

PINS.—Twenty-four thousand one and one half inch standard locust or oak pins, painted with two coats of paint, also end protectors for arm of one inch strap iron, one quarter inch thick to prevent end wire dropping into street in crossing.

BELTING.—Belting to be leather and endless.

If the lights are to be placed upon poles, insert the following:

POLES.—One thousand thirty-feet iron (or wood) ornamental poles with steps.

Or, if over street intersections, insert this:

INTERSECTIONS.—One thousand mast arms, arches or other devices (state which), capable of holding the lamps forty feet high at street intersections.

If towers are desired, insert this*

TOWERS.—One hundred standard one hundred and fifty foot electric light towers, description to be stated, with supports for six lights each; one hundred standard one hundred and fifty foot electric light towers, with supports for four lights each.

PLANT COMPLETE.—The bidder to bid on furnishing the above plant complete in every particular, furnish all labor, including superintendence, and construct and build the circuits, etc., and turn over to the city ready for duty. Bids to be in detail, stating name and kind of each article and price.

MAINTENANCE.—Bidder to accompany bid with a schedule showing the number and approximate salary of labor required and other cost of maintaining the complete plant per year, each item to be carried out in detail. The cost of steam and electrical plants to be kept separate where bidder is estimating on entire specifications.

Where the wires are to be placed underground, the specifications should provide that:

The cables shall have a carrying capacity equal to No. 4, 5 or 6 B. & S. Gauge wire, bidder to give price on each. Said wire shall be thoroughly insulated and character stated. The cables shall have an insulation resistance of not less than ten megohms per mile when laid in the ground. The contractor must estimate on digging all trenches, furnishing all appliances, stating their nature, removing and repairing paving, and laying cables ready for use, guaranteeing them for at least two years.

---

*It should be borne in mind that if towers are used the number of lights necessary will not be so large as by the pole or intersection systems.

INCANDESCENT STREET LIGHTING.

DYNAMOS.—Dynamos capable of generating current for not less than four thousand incandescent lamps of fifty candle power each. The dynamos to set upon insulated bases and provided with belt tighteners. The dynamos will be capable of generating the current as stated for a period of at least twelve hours continuous running daily, without excessive or abnormal heating or sparking at the commutator brushes, or unduly heating the journals. The dynamos will be practically automatic in their action, so as to permit any number of lights to be turned on or off at will without affecting the others in use. The bidder will state class to which his dynamos belong. Also give electrical data of same.

ELECTRICAL APPLIANCES.—Furnish necessary rheostats for setting the candle power of the lamps; same to be of sufficient capacity to permit of the full range of the dynamos, from the burning of the lamps only a dull red, to show that current is being generated, to the full rated candle power of lamps, and to control the E. M. F. of dynamos without unduly heating. The rheostats to be properly connected to the fields of the dynamos by best insulated wire; necessary volt meters, to put in the lamp circuit, and remain continuously in such circuit, to indicate clearly at all times electrical pressure at which lamps are being operated; necessary ampere meters to be also put into the circuit to correctly indicate the number of lamps in operation, and to denote any changes in the number of lamps burning; necessary lightning arresters to be connected into the lamp circuit, same to be provided with proper ground connections; necessary main machine switches to enable the operator in the dynamo room to disconnect the lamp circuits at any time from the dynamos; furnish all other appliances constituting part of the system of the bidder, stating what they are. All of the above to be erected and properly connected in the dynamo room, and made ready to be attached to the power.

POLES, CROSS ARMS AND INSULATORS.—The bidder will furnish all poles, provided with cross arms, pins and insulators, erected ready for the mains to be but upon. And the bidder will put up wires hereinafter mentioned on the poles, furnishing all labor and material of the best quality to properly comply with the requirements.

WIRING.—Furnish a complete system as follows: Mains from station of sufficient conductivity to operate four thousand lamps of fifty candle power distributed over (state number) square miles. The whole to be done so that the loss in electrical pressure shall not exceed five per cent. between the dynamos and the lamps, when the full number of lamps are burning. The bidder to furnish a standard volt meter of well known make. All wires on poles to be well insulated, water, weather and fire proof.

DETAIL WIRING, ETC.—The bidder will furnish wire (state kind) and all line material and labor to properly wire to the street lamps designated by purchaser, and do such wiring in neat and substantial manner.

LAMPS, SOCKETS, ETC.—Furnish four thousand fifty candle power lamps, or equivalent in lamps of other candle powers, and four thousand sockets for same. Furnish four thousand street lamp reflectors with brackets of neat and suitable design, said brackets to be attached to wire poles at height of twenty feet from the ground. The bidder will state at what price he will furnish fifty candle power lamps or equivalent for renewals.

LABOR.—Furnish all labor, including superintendence, connected with the erection of the plant.

POWER.—The plant when finished to be driven by power to be furnished by the bidder as per these specifications.

BELTING.—Belting to be leather and endless.

MISCELLANEOUS.—The bidder will guarantee the plant to work properly and that the lamps will burn steady, and will not blacken inside, and will guarantee the average life of lamp. Also state the number of volts E. M. F. required to properly operate the lamps. Also how many lamps per mechanical horse power at the dynamo the system will produce.

MAINTENANCE.—Bidder to accompany estimate with a schedule showing the number and approximate salary of labor required and other cost of maintaining the complete plant per year, each item to be carried out in detail. The cost of steam and electrical plants to be kept separate where bidder is estimating on entire specifications.

Purchasers of incandescent plants for street lighting purposes should bear in mind the following points in making up their specifications:

SPECIFICATIONS.—For an electric municipal lighting system for large and small cities and villages. Plants from 150 to 10,000 incandescent lights varying from 20 candle power to 100 candle power.

PROPER AMOUNT OF ILLUMINATION.—To be not less than the present gas illumination, uniformly distributed so as to light the crossing and the middle of the block with equal brilliancy, also the alleys, and to give an equal and uniform illumination on the side-walk below shade trees. The central portion of the city to have lamps of 30, 45, 60, or 100 candle power, as the authorities may decide. The alleys and the outskirts may be illuminated by 30, 20 or less candle power as may be designated. The size and the candle power of the lamps may be based on the amount of taxation or the value of the property.

ELECTRIC PLANT.—The size of the dynamos for this purpose may vary according to the size of the city, from 150 to 600 thirty-candle-power each, or its equivalent in any other suitable candle power. None of the above sizes of dynamos to produce more than two distinct circuits. Each of those circuits must be absolutely independent of each other, and perfectly automatic to adjust for any load or any number of lamps on each circuit and to lower the horse power of the engine in proportion. The current generated by either of the dynamos not to be greater in amperage than to require a heavier wire than Nos. 9 or 8 (American Gauge). The loss of current in overcoming the line wire resistance must not exceed one 30 candle power lamp per mile. All the metallic parts of the machines, brushes and regulation to be so arranged and have only such currents of such low potential that both brushes can be handled free of all danger. The automatic regulation must be so arranged that one circuit of the dynamo may be short circuited at the station while the other may do the lighting on the street. Each circuit must be so arranged that it can instantly be connected to any other machine. The dynamo speed must not exceed 850 revolutions per minute. They must be guaranteed to produce not less than seven 30 candle power to the horse power or its equivalent in any other candle power. The switchboard must be provided with an automatic device, which, in case a line or circuit should break and fall on the street or on a tree, will immediately release and disconnect said line from the dynamo and thereby empty the wire of its current. A suitable switch and apparatus must be provided for so that the total power or capacity of the dynamo can be exerted within the station for the purpose of testing the engine under its full load.

POLES AND LINE WIRES.—The poles to be from 25 to 35 feet and from five to six inches in diameter on the top. The same to be properly trimmed, stepped and painted, put not less than four or five feet in the ground and guyed if necessary. Suitable cross arms to be provided to carry the various lines. The line wire to be weatherproof of the best kind with two or three braids thoroughly saturated with weatherproof compound, No. 8 or 9 American gauge hard drawn copper wire. Porcelain loop knobs to be used for the loops down to the street lamps, the very best rubber wire in the market used for entering the street fixtures.

WIRE SYSTEM.—The system to be strictly series so as only to require one single wire on the street from which a direct loop is made to the fixture which holds the lamp. Purchaser must be enabled to insert any candle power of lamps anywhere on said lines without requiring any additional wire or alteration of the same.

THE LAMP HOLDER.—Must include the automatic short circuiter for which absolute guarantee must be given of its closing the circuits under any and all possible conditions as follows: When the lamps naturally burn out; when the lamps are pulled out by hand or otherwise removed, or when the glass is broken; when the filament breaks during the day while there is no current on the lines. For large cities, to insure continuous lighting, each fixture or post should have a double holder with two lamps, one to be the reserve for the other. The same must be so arranged that the continuance of the electric current over the line is absolutely assured, no matter which lamp or filament or globe is broken or removed, and under what condition the breakage takes place, the current being on or off. Even if the second lamp should not be in place and the first one should break, the short circuit must be assured. No fuse or any other contrivance can be admitted that needs replacing after the burning out of the lamp.

THE STREET BRACKET HOOD.—The same must be a substantial structure so as to last under proper care and with proper paint at least fifty years. It must have the highest insulation possible so as to guarantee under any kind of weather or storms uniform lighting throughout the lines, no matter how long they may be. The pole must be made to carry the wires and the street lamp fixture at the same time.

THE LAMPS —The lamps must be guaranteed to last not less than 600 hours on the average. They must be guaranteed not to blacken on the inside, but to stay perfectly clear until burned out. They must be guaranteed to maintain their rated candle power with the original standard of current until the end of their life.

Specifications for isolated indoor incandescent lighting are scarcely necessary in this volume, but as a sample of what ought to be required the following is appended. The plant is supposed to be of 250 light capacity of sixteen candle power lamps.

DYNAMOS.—One dynamo capable of generating current for 250 incandescent lamps of sixteen candle power each. The dynamos to set upon insulated bases and provided with belt tighteners and to be capable of generating the current as stated, for a period of at least twelve hours continuous running daily, without excessive or abnormal heating or sparking at the commutator brushes, or unduly heating the journals. The dynamos will be practically automatic and capable of permitting any number of lights to be turned on or off at will without affecting the others in use.

LAMPS.—Two hundred and fifty lamps of sixteen candle power. Bidder to state price of lamps for renewals.

SOCKETS.—Two hundred and fifty sockets for lamps.

ELECTRICAL APPLIANCES.—Necessary rheostats for setting the candle power of the lamps; same to be of sufficient capacity to control the E. M. F. of dynamos without unduly heating. The rheostats to be connected with dynamos with good insulated copper wire; necessary balancing rheostats for enabling dynamos to be operated in parallel (multiple arc); necessary potential indicators to put in the lamp circuit and remain continually in such circuit, to indicate clearly at all times electrical pressure at which lamps are being operated; necessary ampere meters to be also put into the circuit to correctly indicate the number of lamps in operation, and to denote any changes in the number of lamps burning; necessary ground detectors to show instantly the location of any ground on the circuit; necessary switches and switch boards. All to be erected and properly connected and made ready to be attached to the power provided for.

WIRING.—Furnish a complete system from the dynamo to the main switch or switch board (as the case may be), and from the main switch (or switch board) to the point of distribution, the mains to have sufficient conductivity to operate 250 lamps of sixteen candle power. The wiring to be calculated and put up so that the loss in electrical pressure shall not exceed five per cent. between the dynamos and the lamps, when the full number of lamps are burning. The mains to be of best quality of copper, 98 per cent. conductivity, and of the best insulation. Mains to be put up in good and substantial manner, and have double pole safety devices inserted where leaving the main switch (or switch board).

BRANCH WIRING.—All branches from mains will be attached thereto by means of good soldered joints, and such joints will be thoroughly protected by insulating with some standard compound. Double pole safety device will be inserted at convenient points near the attachment to mains. The branch wires to be of best quality copper. The branches to be so calculated and put up that not to exceed five per cent. shall be lost in overcoming the resistance from the mains. The entire loss of current in mains and branches will not exceed five per cent.

LAMP ATTACHMENTS.—Furnish suitable ceiling rosettes and affix same to the ceiling where lamps are to be suspended, and suspend the lamps from such rosettes by means of insulated flexible copper cord, and properly attach lamps thereto at such height from the floor as will be required, or attach lamps to such fixtures as may be provided by the purchaser.

LABOR —Furnish all necessary labor to fully erect the plant complete.

MISCELLANEOUS.—The bidder will guarantee the plant to work properly and that the lamps will burn steady and not blacken.

MAINTENANCE.—Bidder to accompany estimate with a schedule showing the number and approximate salary of labor required, to run this plant; also the cost of maintaining this plant per year, each item to be carried out in detail.

In dealing with the electrical engineers the purchaser will run against some mysteriously technical words, a great many of which it is not necessary for him to become familiar with. A few, however, would do him no harm to know. The simple ones are these:

The Ampere is the unit of the strength or volume of the current.

The Volt is the unit of the electro motive force or pressure urging the current along.

The Ohm is the unit of resistance opposed to the flow of the current.

The Watt is the unit of power, 746 watts equal one horsepower.

One Volt will force one ampere of current through one ohm' of resistance. Its value is purely arbitrary, but fixed.

The value of an ampere may be defined as that quantity of electricity which flows per second through one ohm of resistance when impelled by one volt of electro motive force.

The higher the resistance opposing a current, the more pressure is required to force the current through.

The various devices which make electric lighting machinery practical, convenient, and economical in operation are these: *Current Indicators* show the number of lamps in use; "*Volt*" or *Pressure Indicators*, show the pressure or "electro-motive force" of the current. *Switch Boards* are for throwing dynamos in and out of circuit. *Gang Switches* are for controlling groups of lights; *Automatic Regulators*, compensate for slight changes of speed; *Safety Devices*, or "*Fusible Plugs*," to cut lamps out of circuit, should there, for any reason, be an abnormal pressure on the circuit; *Wall Plates, Portable Plugs, Hand Rheostats, Swinging Bracket Joints, Shades, Shade Holders* are other special fixtures which render lighting convenient and satisfactory.

There is one aspect for city councils to regard the problem of electric lighting, should they undertake to construct their own plant. It may be considered as practically certain that, whatever be the system put in, patent litigation will be likely to follow. This opinion is based upon the fact that while all systems for conveying electricity are comparatively new, there has yet been a very considerable number of patents granted thereon, covering a wide range of plans, processes, and manufacturing appliances. Nevertheless, "the state of the art," and the validity of what may be regarded as generic patents, have not yet been decisively defined by the courts. Attempts have been made to act upon this subject, and, to be properly passed upon, it has required an extended and painstaking search, involving considerable expense, but at best the result thereof is but an opinion, with which the courts after all do not agree, and yet this is a feature of great importance which may not be disregarded.

Economy in Steam.

THE rapid progress of electric lighting, as a business, has given a new interest to every force, and to every mechanical appliance or device, which bears any relation to the economical and effective installation and operation of electric light plants, whether by municipal authorities, private corporations, or individuals.

The use of steam occupies so important a relation to the production of electric light; and the appliances for the generation and application of its forces form so important a part of an electric light plant, and enter so largely into the question of its economical operation, that no discussion of the subject of electric lighting would be complete without a consideration of the best methods and appliances, and of the latest improvements for the most economical generation and the most advantageous use of steam.

Any device which will economise the production of steam, or increase the profitable uses to which it can be applied, when once generated; or which, in other words, will insure the largest economical results from a given quantity or value of fuel consumed, should receive the closest attention from those engaged in the production and distribution of electric light.

No more important question can claim their attention, than that of making the most profitable use of their exhaust steam, after it leaves the engines or pumps, or of converting it into a source of revenue.

Electric light plants in cities and villages, or in large industrial establishments, are usually centrally located, and surrounded by neighboring buildings, in which, in northern climates, heat is required for the larger half of the year; and, in many cases, all the year round, for certain industrial purposes, such as drying, boiling, dyeing, bleaching, etc.

Exhaust steam, in its normal condition, as it comes from the cylinder of an engine, at a temperature of about 212 degrees, is not, owing to its moist and sluggish condition, an efficient or satisfactory agent, for even the simplest and most ordinary requirements for heating; while for any process requiring a temperature of over 212 degrees, it is useless. By raising its temperature and evaporating its moisture, it may be made to do as valuable and efficient service in heating as live steam direct from the boiler at high pressure.

In this connection attention is called to a system of re-heating exhaust steam, and superheating live steam, without cost for fuel, by the utilization of the waste gases of combustion in their passage from the furnace to the chimney. By this device the temperature of the steam passing through it is raised to within 50 degress of the temperature of the flue in which it is placed, which, under ordinary conditions, varies from 400 degrees to 600 degrees Fahrenheit. At the same time the moisture carried along in the steam is thoroughly evaporated, and the steam is thus rendered dry and elastic.

By this process the temperature of exhaust steam can be raised to from 350 to 500 degrees, and circulated through a properly arranged piping system, without any appreciable back pressure on the engine.

It will readily be seen, that by the use of this system, the exhaust steam from the engines of electric light plants, which is in many instances exhausted into the atmosphere and lost, can be made an important source of revenue in heating surrounding buildings in winter, and in supplying high temperatures at nominal pressure, when required, to neighboring manufacturing establishments.

The same device is also applied to super-heating live steam; raising its temperature to any desired extent without increasing the pressure at which it may be convenient to deliver it from the boiler, and thoroughly evaporating its moisture, and rendering it dryer than it is possible to produce steam in the boiler in contact with water, and capable of doing more effective work in the engine, with less cylinder condensation, than ordinary saturated steam.

In the application of this system to steam plants, it is so arranged, by means of the proper connections and valves, that either live steam from the boiler, or exhaust steam from the engine or pumps, or both together, can be passed through the apparatus at will. In this way, when heat is required while the engine or pumps are not running, live steam at low pressure can be heated up to the temperature due to high pressure, and used for the heating requirements until exhaust steam is again supplied. Also, if at times there is not sufficient exhaust steam to meet the requirements for heating, it can be utilized for the purpose as far as it will go, and live steam can be supplied from the boiler to the re-heater, in combination with the exhaust to make up the deficiency.

By the application of this system to municipal electric light plants, the exhaust steam from the engines employed for running the dynamos, can be utilized for heating the public buildings, and save the fuel that would otherwise be consumed in generating separate steam for that purpose.

Another important feature of municipal legislation of late years has been the proper disposition of the city garbage. It has been decided by the best informed sanitary engineers that cremation affords the readiest and most healthful means for disposing of this necessary surplus. In accordance with this idea various garbage burning furnaces of different designs have been invented and are now in successful operation in Montreal, Pittsburg, Des Moines and Chicago. Other cities are putting them in.

The connection these furnaces have with electric lighting is this: In putting in furnaces in a prospective plant, the city might profitably put in enough garbage burning furnaces to dispose of all the garbage, and thus utilize the heat under the boilers that would be wasted if the garbage cremation was carried on in a separate station. Private electric light companies might also profitably take care of the city garbage by contract and make a saving on their fuel bills.

Importance of Belting.

WHEN the electric light first came into use, electricians found it exceedingly difficult to get a belt suitable to use on dynamos. All belts that were tried seemed to lack something, and even the heaviest and most carefully made leather belts failed to do the required work. This problem received special attention by some of our most enterprising belt makers, and various methods were introduced for making belts specially adapted for use on dynamos. About five years ago experiments were made and a belt was produced particularly fitted for transmitting power when run at a high rate of speed.

The most important operation in the manufacture of dynamo belting, is the stretching. The steer hide is by nature peculiarly formed, having fine and solid fibres on the back, and rounding off towards the belly with loose and coarse fibres; therefore, to get a perfectly flat piece of leather of even tension, and from the center of the hide, it needs the utmost care. The side pieces of the center are generally put into the stretching strain, leaving a short piece free, and thereby stretching the whole piece on an equal strain; or, to explain it more fully, if you wish to stretch a 24-inch piece of leather, you

A STEER HIDE.

place your clamps eight inches on each side and leave the centre eight inches clear; this will draw up enough under the strain of the side pieces to be perfectly even; or rather, the side pieces which contain more stretch, will draw up even with the centre. A dressing or extra coating for the leather was prepared, which made it very pliable and also very smooth, by filling its pores.

The lines drawn inside of the whole hide, show the heart or the most solid parts of the pure oak-tanned leather, which only can be used for making a good leather belt. The bellies and shoulders are finished and sold for shoe purposes.

Another important improvement for dynamo belts, was making them light double, and perfectly even, so that every square inch of the whole belt was of even weight and tension throughout. This caused the belt to run without vibrating like ordinary belts, to run more steady and give the uniformity of power so necessary for producing electric light. Nor did the belt con-

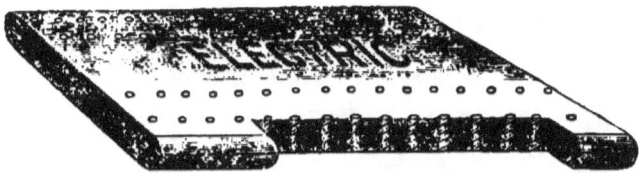

tain copper rivets and burrs or other obstructions, but simply small, endless, wire screws. Its tensile strength was also greatly increased by fastening the edges of the belt with these wire screws, which held the leather firmly together, and yet offered no obstruction to the smooth surface of the belt. These belts met with great success, and were so well calculated to use on dynamos, that they were named "Electric."

Another belt which has lately become very popular, is leather link belting. It was first introduced into this country some two years ago. Since that time so many improvements have been made in this line of belting, that leather link belts can now be used on any kind of machinery. They are made in the following manner: Small pieces of solid selected leather are dressed with tallow and neat's foot oil, which acts as a lubricator to the joints of the pins. The leather is then put

IMPORTANCE OF BELTING.

through rollers, and made very solid; it is then cut into small links; this process makes a link of remarkable tenacity and strength, and one which will stand more strain than a piece of hard rolled *sole leather*. The links are then carefully assorted as to thickness, and the belt built up to the required width. It is of great importance that each belt be made up of accurately assorted links, in order that an even width and perfect running belt may be secured. A patent has been granted for this process, dated January 31st, 1888.

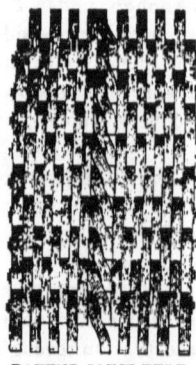
PATENT JOINT BELT.

The Patent Joint is a great improvement in this line of belting. In this patent two bolts are used to each width of belt, and the two widths ingeniously joined together so as to form an unbroken surface. This makes a belt which will conform itself to any pulley, whether flat, rounded, or cone. Other patents have been granted for this style of belt, but they are of minor importance. Link belts should be run as loosely as possible and

IMPORTANCE OF BELTING.

should not be taken up unless they actually slip. An inexperienced person upon seeing one of these link belts running on a dynamo, would naturally think that the belt was loose and needed tightening, but these belts have such a remarkable grip power, that though when running the upper side of the belt is so loose that it almost describes a semi-circle, the under side is as tight as possible.

Before starting new belting examine your shafting and pulleys and be sure that they are true and in line. All pulleys should be crowning in the centre, except those used for shifting belts. Have your pulleys as large in diameter as possible, also for your guide pulleys. Get your belts an inch wider than necessary; it will prove a very profitable investment. An overworked and overstrained belt is the most expensive article; it causes much trouble and delay. An easy running belt keeps the machinery in constant motion. Keep your belts clean; they will last much longer.

How to Light a City.

JESSE M. SMITH, of Detroit, Mich., an electrical engineer of large experience, furnishes the writer with the appended ideas upon municipal lighting:

The system for electric lighting best adapted for a city, town or village must be determined from the circumstances of the case. It is evident that what would be suited for a city would be entirely out of the question in a village, and even cities with the same number of inhabitants require oftentimes very different systems. A city with narrow streets, compactly built up, calls for a different system of lighting from one with broad streets, bordered with trees and covering a large area.

Electric municipal lighting may be divided into three systems:

1st. Arc lights on high towers.
2d. Arc lights on low supports.
3d. Incandescent lights.

There is little doubt that, except in large cities built of solid blocks of high buildings, a satisfactory lighting can be obtained by the tower system with a smaller number of lamps than by any other.

By satisfactory is meant, not a brilliant lighting of the centre of the city with the residence portion and outskirts in darkness,—but a general and nearly uniform lighting of the whole area sufficiently to enable persons to walk and drive comfortably. The most perfect tower system would be one in which each tower supported one large powerful arc light of 2000 or 4000 nominal candle power. This would be manifestly very expensive in towers and attendance, so that very good practical results are obtained by a grouping of three or four such lights on each tower.

It will rarely happen that the tower system can be used exclusively in any place. There will always be dark corners that cannot be reached except at great expense, and in such localities a single low light is used with good effect.

The advantages of the tower system are only realized when the lamps are placed high enough above the trees and buildings, that they may radiate their light without obstructions *near the lamps*.

The lamps should not be at so great an elevation, as that their light will be dissipated in the air before reaching the ground. Towers 75 to 150 feet high have given good results.

Lights of less than 2000 nominal candle power will probably prove unsatisfactory if placed on towers. One of the principal advantages of the tower system is the absence of those intense and sharply cut shadows which are noticeable when the arc lights are placed near the ground.

If it is desired to have gay and brilliantly illuminated streets, no system can compare with that of arc lights on low supports. The supports may be posts set on the curb line; arms extending out beyond the curb, or arches or wires holding the lamps in the center of the streets. Post lights are of little value on streets that are thickly shaded by trees. A small obstruction *near a lamp* shuts off the light from a large area.

On shaded streets mast arms extending out beyond the trees should be used, or the lamps should be suspended in the center of the street. To obtain a brilliant illumination a lamp will be needed at each street crossing and also midway between. These lamps should be placed not less than 30 feet from the ground, otherwise the glare of the light will be disagreeable to persons riding or driving toward them.

Lamps of more than 2000 nominal candle power give more light than can be used to advantage at one point if placed on *low supports*. It will sometimes be found desirable to use "half arc lamps" which give about 1200 nominal candle power. These require about half as much power as the full arc lamp of 2000 candles, so that more of them may be used and a much better distribution obtained.

The electrical apparatus for a plant of half arc lights will cost somewhat less than one for the same number of full arcs. The cost of erection will be about the same in both.

The boilers and engines need only develop a little over half the power for half arcs, hence their first cost and fuel bill will be about half what they would be for full arcs. The cost of carbon points will not be very different in either case. In places where fuel is cheap the total cost, including interest, of operating a plant of half arc lamps will not be very different from that of a plant of full arcs—and the quantity of light will be much less.

Very little has been done with incandescent lamps for street lighting until within the past year. The principal reason for this, has been that incandescent lights could not be carried to great distances without great expense for conductors. This difficulty has been overcome by most of the principal electrical companies, so that to-day the incandescent light may be carried practically to any distance that arc lights can.

Incandescent lamps for long distances, if the direct current is used, are connected by the multiple series system of distribution, in which case each lamp is supplied with an automatic cut-out so that if it fails or is turned out, it will not affect the other lamps. If the alternating system, which is now so prominently before the public, is used,—each lamp is entirely independent of all others. The same *quality* of light may be obtained by either system.

If the quantity of light produced by an arc lamp be compared to that produced by incandescent lamps for the same expenditure of power, the latter appear at great disadvantage. The power required for one arc lamp of 2,000 *nominal* candle power will produce eight incandescent lights of 16 *actual* candle power.

The light produced by an arc lamp is very irregular in quantity and quality and is far below its nominal rating. This is due to the wearing away of the carbon points and the constantly changing length of the arc, even if the machinery in the station is kept in its proper and normal condition.

The light from incandescent lamps under the same condition is practically constant as to quantity and quality. The lamps depreciate after a time and the lights change, but this is not appreciable with good lamps under three months' use.

An arc lamp gives more light than can be used to advantage at one point unless a brilliant illumination is required.

All lights lose their effect very rapidly as you recede from them. The loss varies as the square of the distance. If a light of 800 candle power were placed in the center of a certain area the center would be brilliantly and the outer edges dimly lighted. If this same light were divided into eight lights of 100 candles each and equally distributed over the same area—the illumination would be much more uniform and much greater, although the same total candle power is used in each case.

If now eight lamps of 16 candle power be placed at the same points, the distribution will be the same as before but the quantity of light will be less in the proportion of 16 to 100. The advantage of the greater candle power of the arc light is therefore off-set by the much better distribution of the incandescent, so that all things considered the incandescent is not at such a disadvantage as would appear from the mere comparison of the candle power of the different lights.

The incandescent system is well adapted for places which do not care to pay for the most brilliant illumination, but will be satisfied with less light providing it is well distributed. This system is well adapted for villages which have never had street lighting, for the reason that for the amount of light needed, the first cost of the plant and its maintenance will be less than for arc lights.

Where water power can be had, it will furnish a very cheap, reliable and satisfactory light, particularly if operated in connection with water distribution. This system may also be used to advantage to displace naptha or gas lights.

A sixteen candle incandescent lamp can be made to give 16 *actual* candles of light under all conditions of weather. Wind, snow, sleet, rain, heat, nor cold has any effect on it. It will give from two or three times as much light as is given by the average naptha lamps in actual practice. It will give more light than a gas burner, burning five feet of the best quality, of gas per hour, and much more than the usual gas lights found on our streets.

If lights of more than 16 candle power are needed at any point, they may be had of 25, 32, 50, 64, 100 candle power, but it must be remembered that as the candle power increases, the power required and the capacity of the dynamos and conductors must be increased in about the same proportion. That is, 50 thirty-two candle lamps may be operated by the same dynamo, etc., required by 100 sixteen candle lamps. The first cost of erection will be less for the former than for the latter. The lights will be twice as large but will need to be twice as far apart to light the same length of street.

The cost of municipal lighting must of necessity be very different in different localities, depending on cost of superintendence, labor, power, carbons, value of money, etc. In order to form some idea of the cost of electric lighting in large cities let us suppose the special case of a city requiring 1,000 arc lamps of 2,000 nominal candle power, to be lighted every night in the year from one hour after sunset to one hour before sunrise, or about 3,700 hours.

The plant will be supplied with 50 light dynamos with a capacity of 1,200 lights. The power will be furnished by twelve high speed automatic engines without condensers, belted direct to the dynamos; each engine driving two dynamos. Steam will be supplied by boilers of 1,200 lights capacity and the whole plant erected and operated in the most substantial and best manner.

Non-condensing engines have been selected because all cities have not an unlimited supply of cheap water. High speed direct connected engines are chosen not because they are economical in fuel as compared to large slow running engines of the Corliss type, but because they have certain other advantages as to space, division of power, steadiness of revolution, etc.

The wires will be placed underground in the center of the city and on poles in other parts. The lamps will be carried partly on towers and partly on low supports as necessity requires.

Such a plant, including land, buildings, power, dynamos, wiring and lamps complete and running will cost, in round figures $350,000.

Each lamp if burned all night will require three 12-inch carbons, or 1,100 per year. The engines selected will consume three and one-half pounds of bituminous coal per horse power per hour, including waste. 1,000 lights require 1,000 horse power, which if used 3,700 hours per year, will consume $1,000 \times 3,700 \times 3\tfrac{1}{2} = 12,950,000$ pounds of coal, or about 6,500 tons.

The cost of maintaining 1,000 arc lights for 3,700 hours per year may be made up as follows:

Management	$ 3,500	
Chief engineer	1,800	
Chief electrical man	1,500	
Two Assistant Engineers at $750	1,500	
Two Assistant Dynamo men at $750	1,500	
Two Firemen at $750	1,500	
Fifteen Trimmers and Linemen at $720	10,800	
Various labor	2,900	
		$ 25,000
1,100 M. Carbons at $12	13,200	
6,500 tons coal at $2.50	16,250	
Oil	1,000	
Globes	1,000	
Supplies	1,550	
		33,000
Interest on $350,000 at 4 per cent	14,000	
Depreciation on $200,000 at 10 per cent	20,000	
		34,000
		$ 92,000

On this basis each light will cost $92 per year, or 2½ cents per hour.

The cost per light per hour will vary from these figures considerably under different conditions.

If a large supply of cheap water can be had, compound condensing engines may be used. Such engines consume two pounds of coal per horse power per hour, if engines of say 200 horse power are used.

This item alone effects a saving of 2,800 tons of coal a year, or $7,000, which brings the cost per lamp per year down to $85.

If the lights are only used part of the night, or by moon schedule, the cost per lamp per hour will be higher than if all night lights are used; because the plant has less hours per day in which to earn money, while, what may be called the dead expenses, such as interest and management, continue night and day.

Large plants are generally run with less expense per lamp per hour than small ones, but small plants are only used in small towns, where the cost of ground is nominal, labor is cheap, wire may be strung on poles, and other advantages make the first cost less, and hence the interest account less.

A plant to be used exclusively for municipal lighting will be quite different from one used for commercial lighting.

In the former all the lights are started at one time and stopped at one time.

In the latter the lights are started at irregular hours, and stopped at much more irregular hours. Some will burn all night and a few all day.

The machinery will be taxed to its utmost from about 5 to 10 P. M. and have very little to do the rest of the twenty-four hours.

A commercial plant requires two sets of men in the station, one for the day and one for the night run.

A municipal plant only runs at night, so that one set of men can do all the work with a little extra assistance probably during the longest nights of winter.

The selection of the kind of engine to be used in a municipal plant is much simpler than in a commercial plant.

As the engines always have the same load, advantage can be taken of using large engines, which are more economical than small ones, when both are doing their proper work.

If condensers are applied to these engines and they are compounded the greatest economy can be obtained.

No plant of more than 50 lights should depend upon one engine.

A 1,000-light plant with five engines of 200 horse power would be very economical one as to fuel and attendance and still be flexible enough to meet emergencies.

Whether large or small engines are used, no municipality or company can afford to use any but the very best.

Great care must be taken to obtain close regulation of speed, heavy flywheels, and as great a number of revolutions per minute as the engine will stand with safety.

The quality of the light depends as much on the steady and regular speed of the engine as upon the electrical apparatus.

It will not pay to manufacture light with poor machinery any more than it will pay to manufacture cotton cloth or flour with a half-built mill.

Good machinery, well put in, is reliable, and costs the least to maintain.

Do not buy engines that are *too large* or *too small* for their work.

The Different Systems.

ALL known systems of electric arc lighting are built upon the same general principle. There is, however, a very wide difference in the results which they accomplish. Cost of construction, economy of power and maintenance, durability, safety of the armatures from burning, and simplicity of action, are all items of great interest to the purchaser. Steadiness of light, compactness of apparatus, protection of the armature and field magnets, as well as perfect automatic regulation, are also very important points which should be considered by everyone looking for the best results in an arc light plant.

There are many details of mechanical and electrical construction which must be combined to produce a perfect and complete system. The dynamo is, of course, the groundwork and source of the vital force of all electric illuminators. To do satisfactory work, it must be thorough and reliable in its action, and complete in all its parts.

In considering the comparative value of the different systems the vital points are: 1st, first cost; 2d, economy, efficiency and depreciation; 3d, reliability; 4th, variety and value of the possible sources of revenue, if the purchaser is a private party, and the sources of economy, if a municipality is purchasing; 5th, safety to life; 6th, effects upon existing property. Some of the leading systems are here given with illustrations showing the different appearance of the dynamos and the lamps:

THE AMERICAN.

The dynamo machines of the American system are the fruit of many years' experience and study by Mr. James J. Wood, the electrician of the American Company. The merits claimed for the American system are that its apparatus is simply and perfectly constructed, occupies little space and gives a great amount of energy with the expenditure of a given power. It is claimed to be substantially built and is operated with little difficulty.

The armature is of different construction from any other, being made in the form of a ring of soft iron wire. It is then

THE AMERICAN DYNAMO.

closely covered with coils of carefully insulated copper wire. These coils are so insulated and placed in the new armature as to prevent the possibility of a short circuit between them.

The commutator plays an important part in the protection of the armature. The narrow copper plates of which it is made are insulated from each other with fire-proof material. It is so constructed that one or more of its sections may be readily removed, without interfering with the remaining ones. The center is composed of a gun metal spider, so constructed as to afford ventilation and gives the greatest strength with the least weight. The spider also absorbs any undue heat that may be electrically developed in the ring. The dynamo is placed on a sliding base so that the belt may be tightened or loosened while it is running. Mr. Wood also claims that his current regulator is different from the automatic regulators of other systems. As the number of lights is increased or lessened the brushes are automatically shifted from the maximum to the minimum point, or *vice versa*, with a corresponding increase or diminution of power. In other regulators the current does the work of shifting the brushes.

The different styles of lamps of the American system are unique in design. The feed rod or rods are governed by a small train of clock work, which gives them precision of movement and prevents the carbons from slipping past each other. There are no clutches or glycerine used in their construction or operation. The frame work is water-tight and insulated from the circuit. The lamp requires no hood to protect it from the

THE AMERICAN LAMP.

weather. The wires enter the binding posts from the top and not from the sides. A peculiar device prevents turning on the current before the lamp is ready for lighting, and so protects it from the chief danger of burning out.

Mr. Wood also has a cut-out and relighter, by which, if the light is extinguished by any cause or from any accident, the carbons are automatically brought into contact and immediately re-lighted.

THE BALL.

The Ball dynamo is made from wrought iron, a feature claimed to belong to this system alone. The commutator and means for driving the commutator and armature are constructed of gun metal, and the mechanical details are so constructed as to form no closed loops for generation of waste or Foucault currents, a feature also peculiar to the Ball machine. This system uses the Gramme armature, pure and simple—an endless iron ring entirely surrounded and covered by an endless coil of insulated copper wire. The machine has two armatures, each taking one-half the tension.

THE BRUSH.

Efficiency, durability and simplicity are the advantages claimed by the manufacturers of the Brush system. The Brush light is really the pioneer in the American field, and new inventions of Mr. Brush have largely increased the advantages of an early start.

The simplicity of the machine, the mechanical work, the arrangement of the pole-pieces of the magnets, the small clearance of the armature, and the theory and working of the commutator are emphasized by its users. The commutators are a special feature, and the mechanical as well as the electrical details are thought out and constructed with care.

As now constructed, a "combination" can be attached to the larger Brush machines, by which the electro-motive force of its current can be immediately changed. Thus a machine running thirty lights of 2,000 candles each can, in an instant, be altered so as to run fifteen lights of 4,000 candles. The power of a

THE DIFFERENT SYSTEMS. 109

THE BRUSH DYNAMO.

single light can be immediately doubled by the use of this attachment.

The automatic lamps contain no clock-work or wheel-gearing

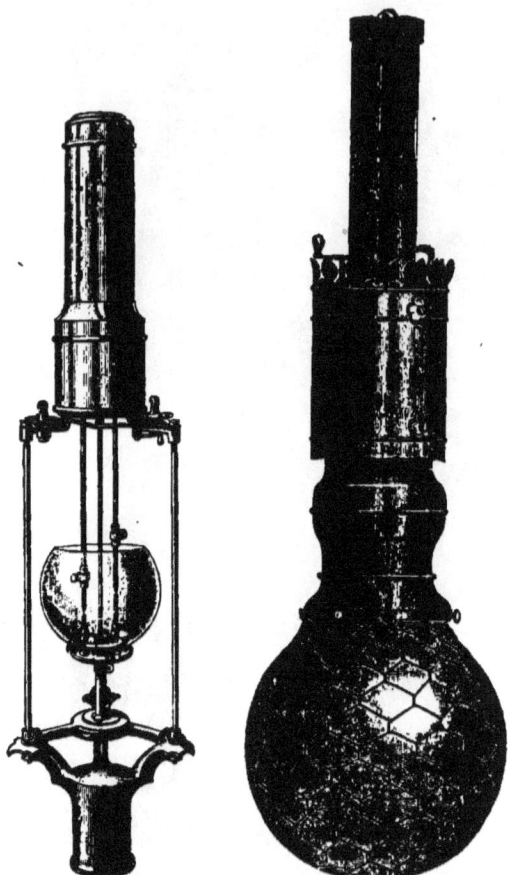

BRUSH ARC LAMPS.

of any sort. The automatic cut-out is a short-circuiting device, by which, when from any accident a lamp is damaged and fails to regulate properly, the current passes uninterruptedly through it and the general circuit remains unaffected.

THE DIFFERENT SYSTEMS. 111

Under the Brush system any number of arc lamps, from 1 to 65 of 2,000 or 1,200 candle-power, or any number from 1 to 20, each of 6,000 or greater candle-power, or a single light of 120,000 candle-power, can be produced from a single dynamo machine. The conducting wires may be extended over circuits of many miles and a hundred lamps or more may be operated on a single wire.

In incandescent lighting the great advantage of the Brush machine claimed is its automatic action in controlling the lamps without resistances, regulators, or other outside devices. The machine is built to operate lamps of sixteen candles, and eight candles each—the sixteen-candle lamp being the standard. Thus the machine for 300 lights will maintain 300 sixteen-candle lamps, 600 eight-candle lamps, 200 sixteen-candle and 200 eight-candle lamps, or any other combined numbers up to its limit, allowing two eight-candle lamps in place of each sixteen-candle lamp. Lamps of higher power, up to 150 candles, can also be used on the same circuit.

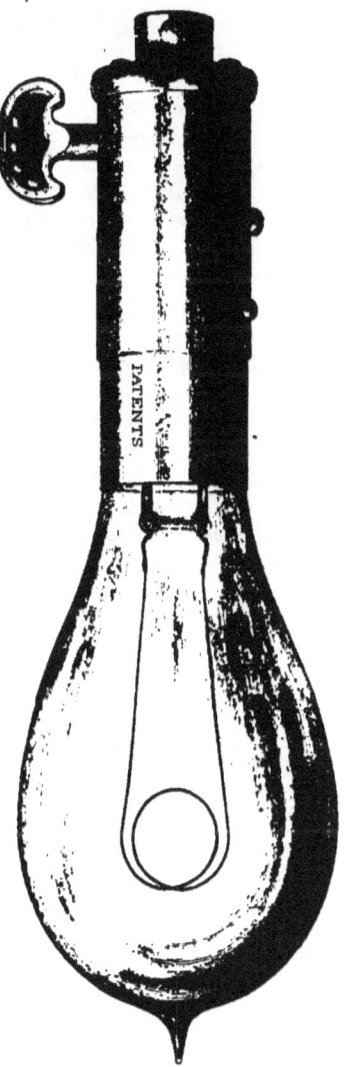
BRUSH-SWAN LAMP.

¶With the dynamo running at a uniform speed, any number

of lights, from one up to the maximum number, may be turned on or off by the automatic regulation of the machine itself. The extinction of a single light shows immediately a corresponding saving of power.

One special feature upon which the Brush company lays stress, is that the success of electric lighting has been due largely to Mr. Brush's researches and genius. The principles which have been recognized as essential, and their practical embodiment in mechanism, are, they claim, of his discovery and invention. These inventions have been covered by broad foundation patents, but many of them are claimed to have been infringed, and a number of suits at law against infringers are now undergoing prosecution in the courts. Doubtless the most important of these is the suit on the double carbon lamp, which has been argued at considerable length in the United States court at Indianapolis before Judge Walter Q. Gresham. It is expected that this suit will determine the entire question of the right to use double lamps, and its result will be of interest, not alone to electric light companies, but to every purchaser of electric light material, be it a private or municipal corporation. This question should not be ignored by the city authorities who contemplate the purchase of plants.

THE CLARK.

The Gramme system of dynamo has been largely used for the foundation work of many machines. The peculiar feature of the Gramme system consists in the style of armature, which is a hollow iron cylinder or drum open at both ends, which revolves around its axis, and on which coils of insulated copper wire are wound in a direction parallel to this axis, which wire is wound in such a way that half of it is on the outside and the other half on the inside of the drum. This drum or armature revolves between the poles of powerful stationary electro magnets, which constitute what is called the magnetic field. The operation of this as well as of many other modern dynamos consists in the fact discovered by Ampere some sixty years ago, namely, that when a conducting wire is moved across a magnetic field an electric current is generated in this wire, the strength of which is increased when the wire is duplicated upon itself by coiling. Another important feature this system has claimed is that the parts of the iron drum which pass along the poles of the field magnets become themselves also strongly magnetized by their influence, which magnetization is a powerful additional agent increasing the electric currents developed.

As this armature is revolving between the two poles of the magnetic field, it was said that only half of the wire in the coils could come under their magnetic influence, namely, only

THE CLARK DYNAMO.

that on the outside of the drum; the other half being situated on the inside could not be reached by the action of the field magnets.

Mr. E. P. Clark, of Owego, N. Y., introduces inside the drum or armature the two pole pieces of a second electro-magnet, so as to have, in addition to the exterior field magnets, also a pair of pole pieces interiorly situated, and which, therefore, can act upon that half of the armature coils which is on the inside of the drum, and in the ordinary construction beyond the influence of the field magnets.

Mr. Clark claims two merits for his machine; first, that it is excited by a very small power and, therefore, that, second, it produces a greater current for the same outlay.

THE EXCELSIOR.

The dynamo of the Excelsior system is the result of careful study by William Hochausen, the electrician of the company. The armature core of the machine is sectional, making it possible to equip it with wire-helices after their insulation has been tested. A small rotary electric motor is employed in conjunction with the Gramme ring.

The current from the dynamo passes through the magnet helices of a controlling device, which sends a portion of this current through the motor-armature, while the balance of it traverses an artificial resistance consisting of two pieces of arc-light carbon.

If the current tends to increase, the armature of the controlling device is drawn towards its magnet till it touches the lower contact-point, sending the current through the motor-armature in such a direction that it revolves the ring forward, cutting field-coils out. When the current decreases, the motor is caused to revolve in the opposite direction, cutting coils in.

The commutator bars are fastened to a stone plate and are separated from each other by air spaces. The current from the machine is sent through the lamps, or withdrawn from them, by means of a switch, which does not break the circuit suddenly, but merely deprives the magnet-helices of the exiting current, lowering

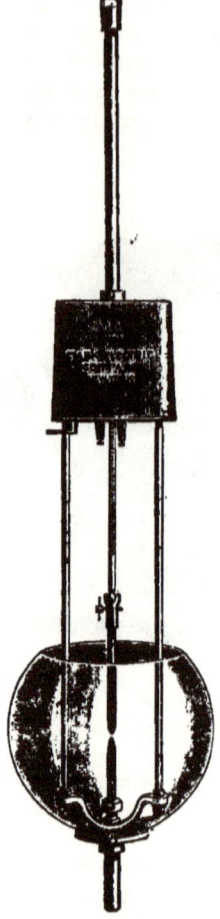

CLARK LAMP.

the power of the magnetic field down to zero and causing a cessation of current in the armature.

The regulation of the lamp is effected in the following man-

EXCELSIOR DYNAMO.

ner: The movement of the upper carbon holder is controlled by a train of wheels, carried on a lever which swings on a fulcrum. The escapement is arrested when the lever is swung so as to lift the carbons apart, and set free when they are caused to approach each other. The end of the lever carries a U shaped iron core, whose straight parts are surrounded by fine wire-helices fastened to the floor of the lamp-case, and has attached to it a retractile spring capable of adjustment. The iron core of a coarse wire-helix is resting on the same lever, and depresses it, owing to its weight overcoming the pull of the spring. This helix forms part of the light circuit, and raises its core as soon as the current is sent through the lamp, thereby allowing the spring to lift the carbons apart by means of the lever and gear-train. When the carbons burn with a small separation the resistance in the light circuit is low; but, as the carbon ends are consumed by the current, the separation increases, and the current has more resistance to overcome in the arc. A shunt receives the constantly increasing amount of current, and draws its core and the lever attached to it, down, till the escapement is released and the wheel-train allowed to move sufficiently to let the carbons approach. No external part of the lamp is in contact with the circuit. To receive a shock while handling it is therefore impossible.

The company claim for the system safety, workmanship and material and durability.

EXCELSIOR LAMP.

THE FORT WAYNE JENNEY.

The dynamo of the Fort Wayne Jenney system was first patented in 1882, and improvements in details have been made from time to time.

Looking at the cut of the dynamo, it will be seen that the magnets consist of two long, heavy pieces of cast iron, bolted together at the point which is neutral or of least magnetic force. The dynamo has but two pieces in its frame, as described, and the advantages resulting from this construction are said to be in

THE DIFFERENT SYSTEMS. 117

FORT WAYNE JENNEY DYNAMO.

the electrical efficiency of the machine and the stability and durability secured by its compact and self-contained form. The concave surfaces of the poles of the magnets are accurately bored out so that the space between the armature and the magnet is reduced to the smallest practical limit. The shaft, which carries the revolving armature, runs in bearings which are cast on the bell-shaped brackets shown in the cut. These are so securely bolted to the frame of the machine as to form, in effect, one piece. By this construction the permanency of adjustment of all the movable parts of the dynamo, is made secure.

The armature is of cylindrical shape, and in form and construction all parts are made easy of access. If any coil of wire be injured so that it is necessary to replace it, this may be done without disturbing any other coil.

The case of the lamp contains a regulating mechanism with a double clutch feed, which controls the feeding of the carbon-rod. This mechanism permits the carbons to come very close together, thus producing a short arc. Each lamp is provided with a switch for extinguishing or lighting it, and an automatic cut-out whereby the lamp is protected in case of accident, as it is immediately cut out of circuit and the current allowed to pass around the lamp.

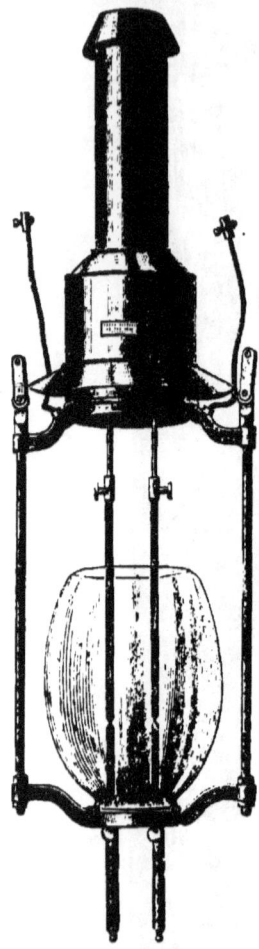

FORT WAYNE JENNEY LAMP.

Believing that the best results are obtained by running incandescent lamps off dynamos made specially for the purpose, this company have made a new machine which is adapted to this purpose. Any or all the lamps can be turned out, and the dynamo accommodates itself to the work without the use of rheostats or other controllers.

THE HEISLER.

The Heisler long distance incandescent system furnishes a five ampere current which is generated and taken from the

THE HEISLER DYNAMO.

stationary part of the dynamo instead of the rotating part, (which is different from the machines of all other incandescent systems.) This stationary part consists of segments which can be disconnected and taken out at any time, thereby enabling the engineer to keep them clean which will insure permanent regular service without repair or expense. The Heisler company lay stress upon the claim that they are combining the llumination of streets, stores and dwellings with 20, 30, 45 and 60-candle lamps on a single wire, No. 8, American guage, hard-drawn, braided, weather-proof covering for outside lines and No. 11 fire proof for indoor purposes. The lamps are connected in series, and no mathematical calculation is necessary. The current can be carried to any desired distance.

HEISLER LAMP.

The first plant of the Heisler system for street illumination was erected in Vincennes, Ind., about the first of October, 1886. One hundred intersections of streets in the suburbs were provided with one 30 candle power light each and supplied by one circuit of 11 miles of wire. The electric illumination in these suburbs is furnished at one-third the price paid for gas in the interior of the city.

Incandescent illumination of the streets on a larger or smaller scale is furnished in connection with the following central station plants of the Heisler system: Monticello, Minn., Mankato, Minn., Pendleton, Ore., Liberty, Mo., Matteawan, N. Y., 170 street lights of 20 and 30 candle power at the rate of $20 and $25 each per annum, distributed with reference to the taxes paid by the property holders, over 18 miles of circuit. Eugene City, Ore., East Portland, Ore., Red Bank, N. J., 75 street lights. Wabash, Ind., 130 30-candle power lights, one at each intersection, a circuit of 12 miles of wire to take place of tower lights. Ocean Grove, N. J., 200 incandescent lamps, one at each intersection, for the illumination of the boulevards.

THE HILL.

A new dynamo has recently been designed by W. S. Hill, of Boston, which is said to have some good features. The frame is made of cast-iron, but the magnet cores and pole pieces are of soft wrought-iron, the effect of which is to give nearly or quite as good results as if the whole frame was made of wrought iron, with a very considerable saving of cost.

The armature is of peculiar construction, and is claimed to develop no heat in the iron, the machine being guaranteed to run cool, unless greatly overloaded so as to cause heating of the wires from excess of current.

THE HILL DYNAMO.

The field magnets are shunt-wound for incandescent lighting and give very perfect regulation. It has as yet been made only in small sizes of ten to 100 lights; a machine weighing 280 lbs. gives 50-70 volts, 16-candle power lamps, at 2,000 revolutions; a 10-light machine weighs 125 lbs.

A large machine for arc lighting is in process of construction and will be fitted with a new automatic current regulator, recently patented by Mr. Hill.

The arc lamp of this system is simple in construction and free from delicate mechanism. It is made to take currents of from four to ten amperes. In a new cut-out switch for arc or series circuits the terminals are mounted at either end of a base board, between which is placed a stand carrying two spring levers, with insulated connectors of thin sheet copper. A cam lever operates to depress and close either circuit before the other can be broken; thus, if a loop or circuit of lamps in a building is to be cut into a main circuit, as the cam lever is turned, the copper brush connectors will be pressed down between the inclined surfaces of the double pole terminals of the new circuit before the connection with the main can be broken. A stout spring that surrounds the axis of the spring levers is arranged to raise and open the circuit of both sets of connectors in common, as the handle is turned in either direction. The switch is covered with a neat box with a glass cover, through which the words "ON" and "OFF" can be seen on the handle of the cam lever.

HILL LAMP.

THE INDIANAPOLIS JENNEY.

This system may be placed under three heads: the arc, multiple incandescent, and series incandescent.

The dynamo used in connection with the arc system is series wound, the field magnets being of the compound horse shoe type with consequent poles. The iron of the field magnets is cast in two parts and bolted together at the neutral points where it is scraped to a surface, thus reducing the magnetic resistance due to joints in the magnet frame.

The armature is wound similar to the Gramme ring, but is much longer in proportion to its diameter. Special spiders are provided by which the armature is rigidly secured to the shaft.

The collector or commutator is composed of numerous strips of pure rolled copper, having strips projecting outward and connecting with the armature coils. The sections of the commutator are insulated with mica and all securely clamped and fastened together. On each side of the magnet frame, and extending from one end to the other is an iron arm which supports the boxes within which the armature shaft revolves. These boxes are fitted in a special jig making them interchangeable and self-lining. A very important feature in connection with this dynamo is the automatic regulator, which shifts the brushes to compensate for variations in the resistance of the circuit, and also for variations in the speed of the dynamo. Any number of lights from one to the full capacity of the dynamo can be turned off and on, and the regulator will preserve constant current on the line. As the E. M. F. generated is only sufficient to overcome the resistance of the line and the lamps in operation, the power consumed by the dynamo is about in proportion to the number of lamps in circuit.

The arc lamp in connection with this system is very free from complicated devices. It has the carbon holder rod, a clutch, two sets of hollow coils, known as solenoids, an automatic cut-out, hand-switch, etc.

The clutch used is so constructed that in operation it does not let go of the feeding rod, but merely loosens enough to allow it to slide the desired distance, when further movement is arrested by the controlling magnets within the lamp. The double lamp works in the same manner but in addition to the

THE INDIANAPOLIS JENNEY DYNAMO.

parts used in the single lamp, is a device which maintains an exact adjustment and equal illuminating power of both sets of carbons. It is immaterial which set burns first.

In connection with this system are all the essential accessories, such as switch boards, ampere meters, lightning arrestors, etc.

The dynamo used with the multiple incandescent, is also of the compound horse shoe form, but in the matter of proportioning and constructing is very dissimilar. The two magnet columns are fastened in a vertical position to a heavy base, which supports strong upright castings upon which the journal boxes are secured. These columns are connected at the top by a massive yoke.

The armature of the machine is also a modification of the Gramme. The iron core is composed of numerous thin soft iron plates, all thoroughly insulated, bolted together and securely fastened to the heavy spiders which are keyed to the armature shaft.

There is but one layer of copper wire wound on the core, and the resistance is very low. This machine is shunt wound and not compounded. All the magnetic parts are very heavy and with constant speed the machine is very nearly automatic.

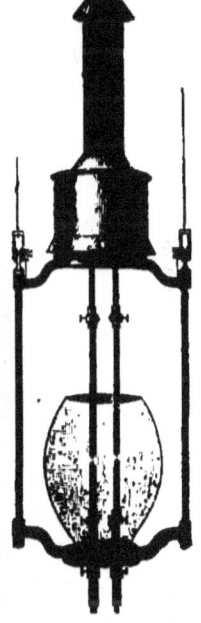

INDIANAPOLIS JENNEY LAMP.

The series system of incandescent lighting is designed to meet the wants of those desiring a system where incandescent lamps can be operated at a great distance from the dynamo with a small percentage of loss in the conducting of wires, and without the enormous expenditure for copper, which would be unavoidable if the multiple system were used. With this system the lamps are placed in series, and may be of different candle power, as no balancing is required to maintain the current at constant strength.

The dynamo used is shunt wound. An automatic regulator controls the current by operating a rheostat which is placed in the field shunt circuit. Therefore, when but a few lights are being used, only a small amount of power is required. All lamps of this system are provided with automatic cut-outs, which short circuit the terminals of any lamp when the carbon in it is broken. The lamps are also provided with a hand-switch for turning on or off, and are so constructed that the lamp cannot be taken from the socket when the switch is turned on, nor can it be switched on when there is no lamp in the socket. It is, therefore, impossible for a careless person to open the circuit and form an arc in the cut-out.

THE LOOMIS.

It is the claim of the Loomis company that their incandescent dynamo is entirely automatic, requiring no more attention

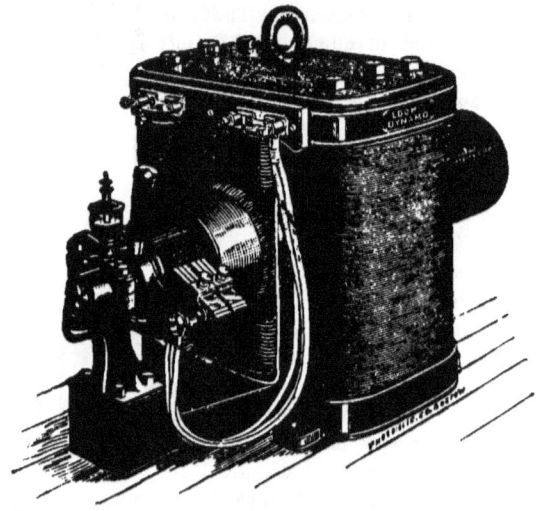

LOOMIS DYNAMO.

than a line of shafting. It is conspicuous by the absence of switches, resistance boxes and other appliances, and, by dispensing with the employment of skilled attendance, diminishes the running expenses. The system is mainly applicable to isolated lighting, but the company furnish an arc lamp which is said to feed its carbons with such steadiness as to abolish all "frying" sounds familiar to arc systems. The company is but lately organized.

THE MATHER.

In form and mechanism the dynamo of the Mather system is unique in appearance and noiseless in operation. It is designed especially for isolated incandescent lighting.

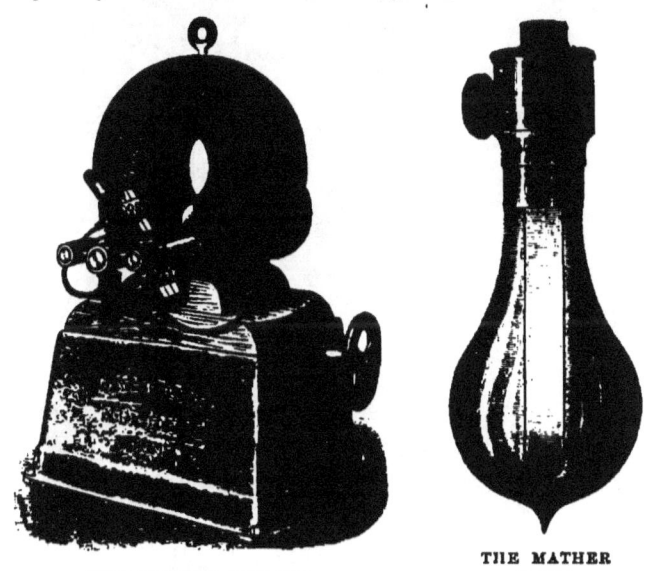

THE MATHER DYNAMO.   THE MATHER LAMP.

The Perkins lamp, which is manufactured by this company,

is claimed to have a characteristic merit of freedom from discoloration. The dynamo is wound for a pressure of 126 volts and produces twelve sixteen-candle power lights to the horsepower.

THE MUTUAL.

The dynamo of the Mutual system is of the Gramme type. Its sectional armature admits of the building of a Gramme ring with little expense, with uniformity of winding, and it is compactly built. It is claimed that the machines embrace

THE MUTUAL DYNAMO.

THE DIFFERENT SYSTEMS.   129

points of novelty and practical utility which tend to make them durable, safe and certain. The lamps work on what is known as the derived circuit principle, there being no differential action whatever; the feed being positive and the length of the arc nearly constant. All of the action is accomplished by the direct action of magnetism and gravity, without springs, dashpots, or similar retarding devices.

MUTUAL LAMP.   SAWYER-MAN LAMP.

A peculiarity of the double lamp is that one set of carbons can be handled and trimmed while the other is in action.

THE SAWYER-MAN.

The dynamos of the Sawyer-Man system are claimed to be automatic in their regulation, and will maintain a uniform light,

THE SAWYER-MAN.

with all or any portion of the lights in circuit. The lamps will not blacken and will maintain their candle power during their guaranteed life.

The Sawyer-Man lamps have a distinctive form of carbon loop, somewhat similar to the many forms of incandescent lamps now before the public. The filament is constructed of the same material which the most successful have used, but the company claim for the lamp superior advantages, resulting from their special method of treating it.

THE SCHUYLER.

The features claimed for the Schuyler dynamo are: Accessibility of every part for inspection and repairs. The dynamo

THE SCHUYLER DYNAMO.

can be taken apart and put together in fifteen minutes. The arrangements for lubricating are peculiar. A Schuyler requires oiling not oftener than once a month. Low speed of dynamo for given output of current.

The series lamp of the Schuyler system is an incandescent or glow lamp which is cut into the arc circuit, without resistance coils, regulator, converter or other auxilliary apparatus. It gives a clear, white light, and is made of any desired candle power.

It is not necessary that the series lamps should be run in groups; it is not necessary that lamps on the same circuit should be of equal candle power. The only limit to the number of lamps that can be placed on a given circuit is the limit of capacity of the dynamo. The series lamp can be mingled indiscriminately with arc lamps upon the same wire, and will furnish about two hundred candles per horse power. The distance to which series lamps can be used from supplying station is the same as with arcs; and the brilliancy or efficiency of the lamps is not affected by distance, nor by the size of the circuit wires, which can remain the same for all ordinary distances,

132   THE DIFFERENT SYSTEMS.

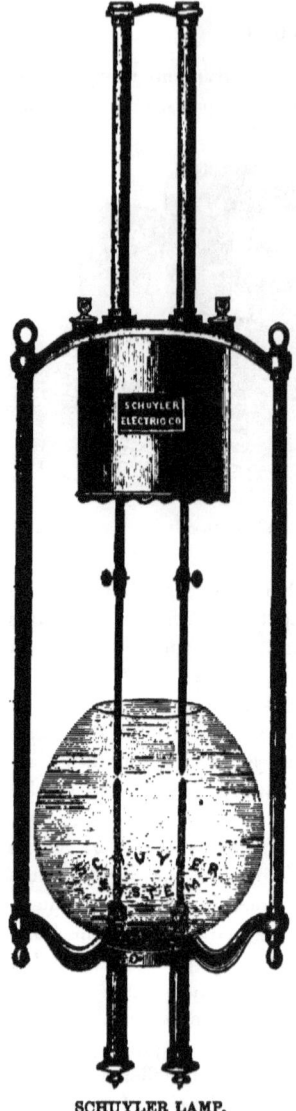

SCHUYLER LAMP.

precisely as for arc lamps. The lamp is furnished with a simple automatic cut-out, within the lamp, to act in case of rupture of filament.

The Schuyler series incandescent lamp can be run on any system using from six to ten ampere current. The Schuyler arc light is generated from a dynamo with a ventilated armature, with which it is claimed to be impossible to have a burn-out.

THE THOMSON-HOUSTON.

In the dynamo of the Thomson-Houston system the field magnets consist of two hollow cylinders, which are supported by a frame, which also carries the bearings for the armature shaft. The inner end of each cylinder is formed by a spherical cap having an opening in the centre, and the two are brought together, so as to leave a spherical space in which the armature revolves. At their outer ends the cylinders are flanged and form annular plates. These plates are connected by iron bars which unite them magnetically, and at the same time act as a protection for the wire wound around the cylinders.

The magnetic field generated by these means is spherical in form and is consequently occupied by an armature similarly shaped. This armature consists of only three bobbins of wire wound around a core of iron resembling in form an ellipsoid of revolution. The ellipsoid is built up of two concave plates fixed on the shaft, and the edges of which are bridged over by light iron cross-pieces. On these cross-pieces there is wound a certain quantity of iron wire which is oxidized and varnished so as to insulate the various layers.

Each bobbin is cut out of circuit the moment it traverses

THE THOMSON-HOUSTON ARC DYNAMO.

the region of small effect, but by means of a simple arrangement, the cutting out of circuit of each bobbin takes place at the instant the latter reaches a point 60° from the neutral line; or, in other words, when it has reached a point at which it is no longer important to gather the current.

Prof. Thomson regulates the current of his machine by varying the position of the brushes with respect to the commutator, and he accomplishes this object in two ways.

The first method, which he terms forward regulation, is employed when a diminution of current is required, and consists simply in advancing the position of the brushes a certain distance on the commutator.

The second method is termed backward regulation, and is that which is now generally applied to this machine. Under normal conditions, the brushes in each pair form an angle of 60° between them, and touch the commutator at points also 60° apart, consequently the angular distance between the first of one pair and the second of the other pair of brushes is 120°, *i. e.*, the length of one segment of the commutator. It is evident that with this arrangement none of the bobbins will be cut out of the circuit, since each segment of the commutator will have to pass from one pair of brushes to the other at the instant the bobbin itself is at the neutral line; thus there are always two bobbins in parallel, joined in series with a third.

The Thomson-Houston machine works at high potentials.

THE UNITED STATES.

The dynamo used in the United States arc system resembles in its general features of construction that used in the incandescent system, but is modified in certain details to produce the different quality of current required for arc lamps. A marked peculiarity of the United States system is the shortness of the arc or separation between the carbons, it being about one thirty-second of an inch in length, as compared with one sixteenth to one eighth of an inch in other systems. This, the company claims, enables a given number of lamps to be worked with a current of correspondingly low tension. Stress is laid upon the high efficiency of the lamps

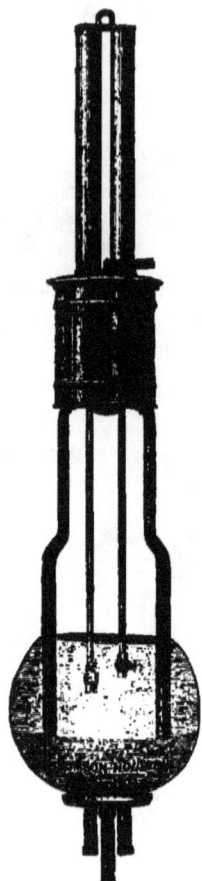

THOMSON-HOUSTON LAMP.

THE DIFFERENT SYSTEMS.

THE UNITED STATES DYNAMO.

and dynamos, which, it is claimed, makes it possible to operate a given number of lamps with smaller boilers and engines than are required with other systems. The company has several different systems of wiring, each of which is adapted to meet certain requirements of station lighting. The company's claims are: Great simplicity, economy of operation, economy in cost of mains, and that it can be distributed widely with commercial success. The company pays particular attention to the incandescent system, both for interior and exterior work.

THE VAN DEPOELE.

The main features of the Van Depoele dynamo consist in the peculiar disposition of the field magnets, the construction of the armature, the sim-

U. S. INCANDESCENT LAMP.

plicity of all parts of the apparatus, compactness, ease of management and control of the current to the work called for.

The field magnets consist of two large coils of copper wire wound around two soft iron cores, their north and south poles facing each other, and between these poles revolves the armature. To the soft-iron cores of the field magnets are cast on one end heavy back plates, while to the other end are secured the semi-circular pole pieces between which the armature revolves in close proximity to the latter. The back plates of the magnets are secured between the top and bottom plates, holding the whole in position and making a most solid frame. To the lower plate, and in its center, which is neutral, are cast two extensions upon which are placed the posts or bearings supporting the armature shaft; this disposition allows the posts to be comparatively short, providing a very rigid support for the revolving armature. Further, the whole frame may be considered as a very long electro-magnet, with its poles inverted toward the center.

UNITED STATES ARC LAMP.

The armature consists of a frame made of a number of iron bars, each separated from the other; these bars are riveted to the inner and outer periphery of two metal rings, several of these rings being placed between the inner and outer layers of iron bars. And finally, the rings and bars are riveted together so as to form a solid frame. It is claimed that with this armature it is unnecessary to provide for ventilation, since it is

THE VAN DEPOELE DYNAMO.

said there is no heat generated. The electro motive force of the current is kept down as low as possible and burns short arcs in the lamps.

The principal features of the lamp are two electro magnets, one in the main and one in the shunt circuit; to the latter is hinged a soft-iron armature, the free end of which moves under the influence of the opposite pole of the main magnet. This armature carries the carbon lifter, so that any motion imparted to the armature, under the influence of its electro-magnets, is directly communicated to the lifter, either separating the carbons or allowing the same to feed. An incandescent lamp is also made by the company.

VAN DEPOELE INCANDESCENT LAMP.

VAN DEPOELE ARC LAMP.

THE WATERHOUSE.

The Waterhouse company claim for their system a full 2,000 candle power light on three-quarters of a horse power, including friction. They also manufacture an eight and one-half ampere light, which they guarantee at .65 horse power per light, including friction. The lamps of this system burn long arc and are made in single carbon lamps either with clutch or rack feed and a double lamp that has the feature of being the same as a single lamp, one side being disconnected while the other side is burning. The lamp magnet is the same in all the lamps and is a new form of magnet. It controls the lifting mechan ism of the lamp so that a steady feed is obtained. The method of regulation for the dynamo is new. If a light or lights are turned out the result on the field magnets is that the current that passes around them is reduced and therefore in cutting out lights down to one light or none, very little current passes around the field magnets, and the machine is not capable of producing

THE WATERHOUSE DYNAMO.

a greater current than the standard. The regulator controls the supply. If the speed lessens, the current increases in the field magnets, and the generating capacity of the machine is increased and the current maintained at standard. Of course, there is a limit to which the speed can be reduced, but the company claim a reduction of 100 revolutions in the speed of the armature will be compensated for by the regulator.

THE DIFFERENT SYSTEMS.

The Waterhouse dynamo is compact in form and has long bearings for the armature shaft. It is of the closed circuit type, with new improvements. The dynamo contains less wire, and in consequence less resistance than some other systems,

WATERHOUSE LAMP.   WESTERN LAMP.

and with regulation, attains great efficiency and saving of power.

The lamp magnet is an iron core, quadrangular in shape, and has the main and shunt circuit coils arranged at an angle from each other. The magnetic effect on the iron core, produced by the currents flowing through these circuits, is the novelty of the invention, and new in its application.

THE WESTERN.

The arc lamp of the Western system differs in internal construction and operation from that of any other. The manufacturers claim that undue complication is avoided, and the working parts are reduced in number, simplified in form, and so arranged as to reduce the liability to derangement from wear, corrosion, dust or other causes, and at the same time to secure regulation of arc resistance and current strength, thereby maintaining a steady and uniform light.

WESTERN DYNAMO.

The Western company also claim that to the high efficiency of their dynamo is due the economy of the system. The dynamo is of compact pattern, with a specially constructed drum armature, composed of but few parts, easily accessible. From the method of winding the armature, economical results, with any number of lights are obtained ; destructive or damaging heat is avoided, and the machine can be run on short circuit (all lamps switched out) without injury. By a simple movement of the brushes backward or forward, any number of lamps, up to the capacity of the machine, may be run without undue sparking at the commutator. This freedom from sparking, it is said, permits the use of oil on the commutator, thus reducing the wear to a minimum.

The field magnets are constructed with a view to securing a seat for bearings and pole plates at the same time. The bearings are plain, and the entire design is simple.

EDISON DYNAMO.

THE EDISON.

The dynamo first constructed for the Edison municipal system gives 1,000 volts, and is made in two sizes, 12 and 30 amperes. The present dynamo gives a maximum E. M. F. of 1,200 volts, which allows for any drop in potential which may be desirable up to 17 per cent. in the wire, while a pressure of 1,000 volts is left to be expended in the lamps.

EDISON LAMP.

The lamp is of low resistance, with thick substantial carbon, the length of the loop determining the candle-power and the E. M. F. required. Hence, as a 15-candle lamp has a carbon of the same cross-section as one of 50 candles, it requires the same current, the difference being simply in the volts absorbed. This gives a flexibility to the system, the only requisite in calculation being that the total candle power in each of the various circuits shall conform approximately to a given standard, which standard is found by a determination of the most economical percentage of loss in the conductor in each particular instance.

The lamps require about four amperes, and have a higher standard of efficiency than the high resistance lamps used in the three-wire system. Their life has been very long, reaching in some cases from 1,500 to 3,000 hours. The standard of distribution for this lamp is 1,000 candles for each circuit of 1,000 volts.

THE DIFFERENT SYSTEMS.

The standard street hood has a metallic frame and top, with an inverted conical reflector of opal glass. It contains a socket and cutout of simple construction, which, in case the safety device in the lamp itself should fail, operates to complete a shunt around the terminals, and also to maintain the continuity of the circuit when a lamp is removed.

An important attachment to every hood when suspended from posts in the open air is an insulator which makes it impossible for the wires, cross-arm or frame to become grounded at this point.

In interior incandescent lighting, to which the Edison company has devoted much attention, the "three wire" system is employed. It is claimed that this system enables the company to use conductors one-third the size that would otherwise be required.

THE WESTINGHOUSE.

The Westinghouse alternating current system is peculiar among the systems.

The first cut shown illustrates the arrangement of the street light on an iron pole, the converter being generally placed at the top of the pole and the wires led down inside.

THE DIFFERENT SYSTEMS.

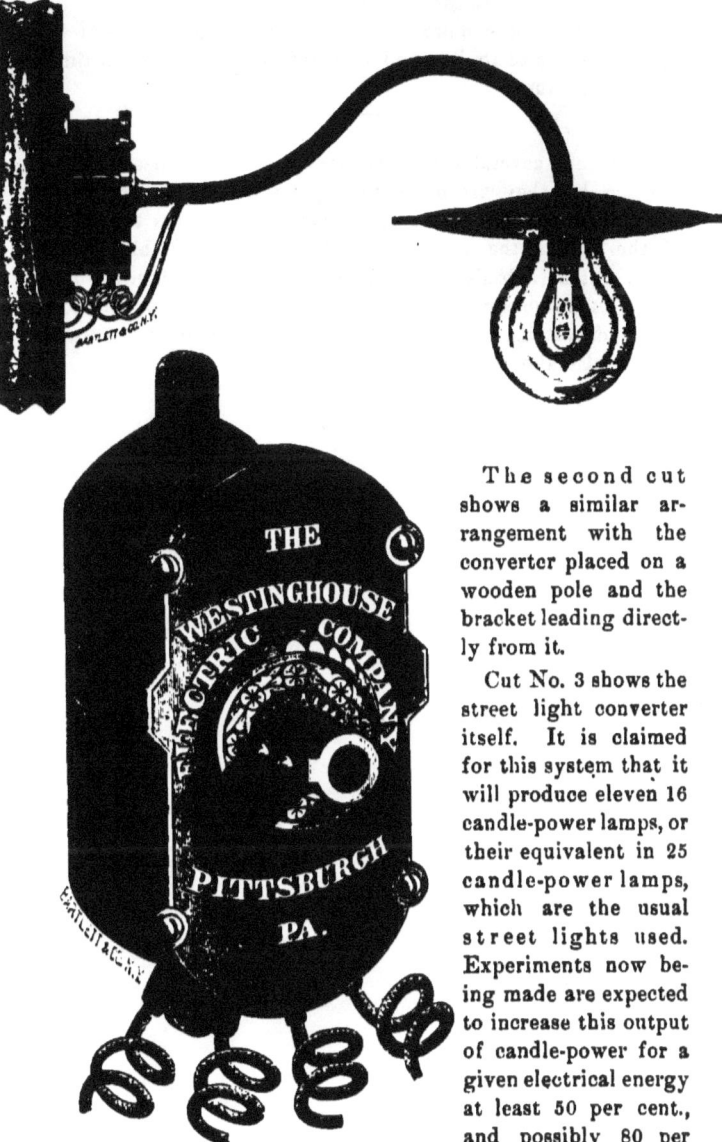

The second cut shows a similar arrangement with the converter placed on a wooden pole and the bracket leading directly from it.

Cut No. 3 shows the street light converter itself. It is claimed for this system that it will produce eleven 16 candle-power lamps, or their equivalent in 25 candle-power lamps, which are the usual street lights used. Experiments now being made are expected to increase this output of candle-power for a given electrical energy at least 50 per cent., and possibly 80 per cent. The Westinghouse company also claims to have been peculiarly successful in their form of lamp, which cannot be made to depreciate more than one candle-power for the 16 in the full length of its life, and this depreciation occurs mainly before breaking.

There are several other systems of arc and incandescent lighting, but they are not generally known or have dropped out of common use. One occasionally runs across the name of the Fuller, the Remington, the Sperry, and this or the other, but as a basis of information they cut no figure whatever. In purchasing, the public or private patron will find plenty of systems to examine without running after unknown or unused makes.

Storage Batteries.

PLANTÉ, the noted French scientist, was the first to demonstrate the possibilities of the "storage battery," or accumulator. The problem of how to store up for use when required electrical energy, generated by whatever means, had been discussed for years, but the most rapid advancement of the system has been made within a very short time.

The introduction of storage batteries has made it possible to obtain certain desirable results hitherto altogether impracticable. A reservoir of electricity is made possible from which light can be obtained when the power developing machinery has stopped; and this reservoir may be located economically at a great distance from the generator. The storage battery is the analogue of the gasometer in gas lighting, acting not only as a reservoir of power, but also as a governor of the current supplied.

The advantages of the accumulator are manifest more particularly in its application to the incandescent form of electric lamps, but the time is not far distant when it will play a very important part in the arc systems. It affords the means of combining the two lights. The same plant which supplies electrical energy to a given number of arc lights (which, because of their powerful character, are, as a rule, best run by direct current), will supply twice as many more incandescent lights with very little extra expense. The current may be supplied to the arc and incandescent lamps by separate wires, the direction of the current being controlled by switches at the station, so that the accumulators for the incandescent lamps can be charged when the arc lights are not in service. The arc lamps for street lighting are fed direct from the dynamo, and the incandescent lamps for house lighting by the accumulators.

Each house after its accumulator is charged is disconnected

from the system, and is therefore independent of all others. No matter what may occur to cause trouble with the street lamps, its lights are secure. Should anything unforeseen prevent the perfect action of the lights in any one house, others are not affected by the disability. In the direct system any derangement to the running machinery affects all the lights in the district.

There is no danger of shocks, serious or otherwise from the accumulator, as the current has not sufficient potency to overcome the resistance of the body. In the direct system there is always more or less danger.

An important attribute of the storage battery is its portability. It may be transported any distance without suffering impairment of its working power. This property confers upon it distinct advantages, opening up entirely new fields for its application. The direct system of lighting limits the use of the electric lamps to the district over which the wires extend, and outside of this territory it is not available. It frequently occurs that a temporary electric light service would be desirable at points remote from regular centers of distribution. Then, the storage battery may be charged at the station, and the temporary installation made at any distance. Any quantity of light desired may thus be supplied at only a few hours' notice, and the quality will not vary from that of lamps on the regular circuits.

Small municipalities, desiring to light the streets with arc lights and the public buildings with incandescent lights, may profitably examine this system.

The batteries are composed of a number of cells containing cast lead plates of a peculiar construction, chemically prepared, and immersed in a solution of sulphuric acid. These cells may be connected together so as to produce any desired result in current or pressure. When placed in a circuit with a dynamo machine, the electricity generated is accumulated in the battery and may be used at pleasure. A large number of these batteries can be placed in one circuit, and be supplied with electricity from one generator.

Where dynamo machines have already been provided for running arc lights, they can be used at any time when not required for the arc lights to charge the storage batteries.

STORAGE BATTERIES. 149

An automatic current manipulator or switch is provided with each battery, and is so arranged as to retain the battery in circuit until it is fully charged, and then disconnect it from the circuit. When the charge has been exhausted down to a certain point, it brings it into circuit again and holds it till it has been recharged, and then cuts it out as before. The same operation is repeated with every battery in circuit. The entire operation is automatic. Each battery has a meter attached, which registers the exact amount of electricity stored.

Distribution of Light.

THERE are three methods of street illumination in vogue in this country—gas, oil and electricity. These are susceptible of sub-division, as, gas into manufactured and natural; oil into kerosene, naptha and gasoline; electricity into arc and incandescent. Manufactured gas is used exclusively in many places where long contracts have not expired, or where the authorities have not been sufficiently enterprising to adopt electricity, or have been unduly prejudiced against it. Natural gas is, of course, in use only where the circumstances of location make it available, and the country contiguous thereto. It is not a good illuminant on account of the intense heat which it emits.

Oil is generally in use in small places where the establishment of gas works would be unprofitable, or the rays of electricity have not penetrated. In many of the large cities oils are still in use in the outlying portions.

Electricity has come to be recognized as the successor of all these out-door illuminants, and its fight for place has not been to any degree more earnest than has the battle between arc and incandescent lighting. The former is the better intrenched, because it antedated the latter, but the progress made by the incandescent light has reached that point where it now unreservedly asserts itself as a competitor, and municipalities should in the future consider both in their discussions. Heretofore the incandescent lamp has been regarded more as an interior light, and its chief competitor was gas.

The entire question of street lighting, aside from the various considerations entering into the adoption of any system, is one of the distribution of light. The advocates of the arc light, the multiple arc and the incandescent will each aver that his particular system will give the best distribution, and he will be prepared to prove it. It will resolve itself in the mind of the average layman like he who "could be happy with either were t'other dear charmer away."

In considering the distribution of light one should not compute the total candle power of a given territory, and raise or lower the lamps until an equal distribution of the light is obtained. There will be alternate dark and light spots in that case. The property owner living in the center of a block is as much entitled to light as he who lives on or near the corner. The effect of light on crime will not be so perceptible if the alleys and back-yards are neglected. To get an equal distribution of light the city should first determine where it wants light and should then study the different systems as a means of putting it there. As someone has said, "the experience of practical users is more valuable in enabling one to determine what to buy, than scientific tests or anybody's guarantee." The attention of the reader is therefore called to that which here follows.

Three methods of placing arc lamps for street lighting are extensively used. These are:—

First.—Placing lamps upon poles.

Second.—Placing lamps over intersection of streets.

Third.—Placing lamps upon towers.

The first method, that of placing lamps upon poles, has but one advantage—cheapness of first cost—while it has many disadvantages; mainly from the fact that the pole can be placed upon one side of the street only, the light is unequally distributed, and, where shade trees exist, only the opposite side of the street can be lighted. A serious disadvantage, and one that deserves more attention than it receives, is the cost of carbon trimming. The trimmer requires time to climb to the lamp, and time is money and climbing dangerous. It will, therefore, be seen that, while the first cost is much cheaper, in the end, pole lighting costs more money than any other.

In some places where pole lights are in use the lamps are simply stuck upon crooked and ill-shaped wooden poles, giving an appearance of shiftlessness that speaks ill of the system and of the company that puts such unsightly objects on the streets. The cut of a street-light here shown illustrates how a pole light can be placed neatly, and with but a slight additional cost over the slip-shod method referred to.

The second method, that of placing lamps over the intersections of streets, is the best method when low arc lighting is desired. It has many decided advantages over the pole method.

By placing a lamp over the intersection of streets, you place the lamp where it can do the most possible good, as it can light the streets equally in four directions a distance of about 400 feet radius of the lamp. The lamp is far enough away from shade trees to allow its rays to penetrate underneath them and thoroughly light the sidewalk. Lamps should be at least thirty-five feet from the surface of the street.

There are four ways by which lamps can be placed in this position. One is to erect two poles upon diagonally opposite corners of the street, and connect them at the top with a twisted wire cable. The lamp is hung on a pulley on the center of this cable; a line is attached and the lamp is drawn in to the top of the pole for trimming.

Another method is that of an iron pulley fastened to the center of a cable, the lamp being lowered to the center of the street. During the extremely cold weather of winter, the cables and cords are liable to become coated with ice and the apparatus fail to work.

DISTRIBUTION OF LIGHT.   153

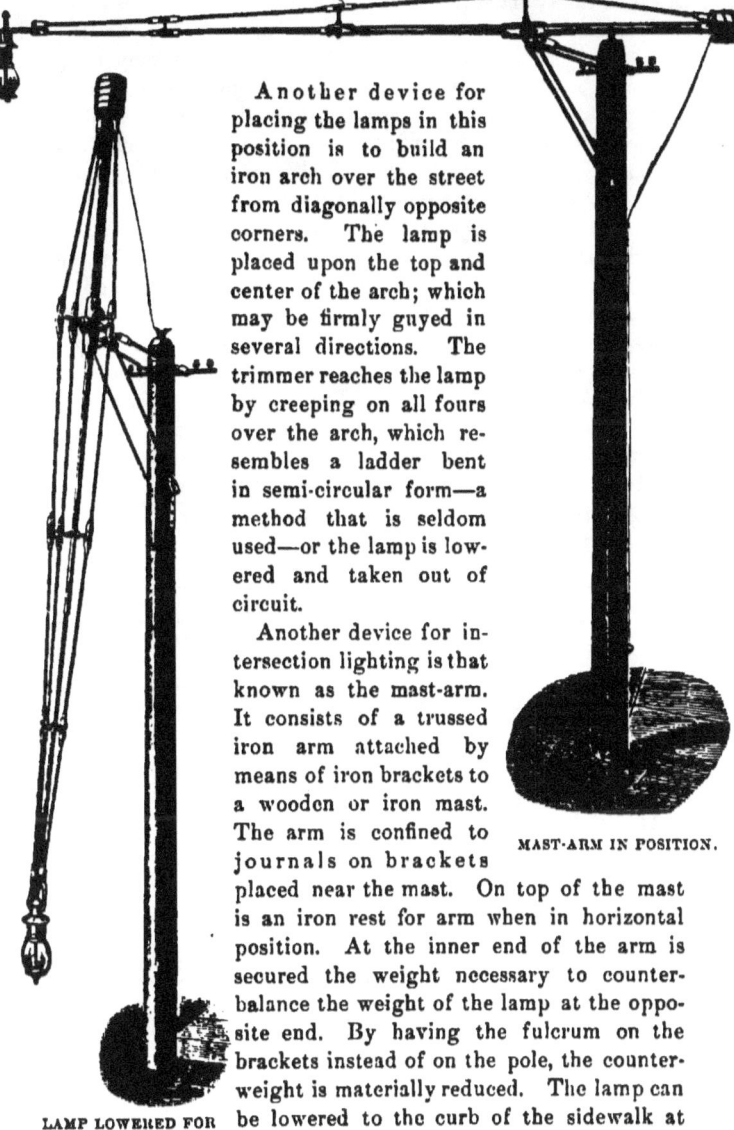

LAMP LOWERED FOR TRIMMING.

MAST-ARM IN POSITION.

Another device for placing the lamps in this position is to build an iron arch over the street from diagonally opposite corners. The lamp is placed upon the top and center of the arch; which may be firmly guyed in several directions. The trimmer reaches the lamp by creeping on all fours over the arch, which resembles a ladder bent in semi-circular form—a method that is seldom used—or the lamp is lowered and taken out of circuit.

Another device for intersection lighting is that known as the mast-arm. It consists of a trussed iron arm attached by means of iron brackets to a wooden or iron mast. The arm is confined to journals on brackets placed near the mast. On top of the mast is an iron rest for arm when in horizontal position. At the inner end of the arm is secured the weight necessary to counterbalance the weight of the lamp at the opposite end. By having the fulcrum on the brackets instead of on the pole, the counterweight is materially reduced. The lamp can be lowered to the curb of the sidewalk at any time for attendance, whether in use or not. In some cases the circuit is opened, in others it is not. Scarcely any exertion is required on the part of the attendant, as the lamp is so balanced, being slightly heavier than the opposite end, as to descend slowly, and a slight strain on the wire line fastened to the inner end of the arm will raise it again in position. An attendant can trim double the number of lamps placed on mast arms than is possible to do in the same time, when lamps are placed on any of the other devices.

Some mast arms may be adjusted to vary the height of the lamp from the ground. The lever bar for raising and lowering the lamp to within five feet of the ground, also acts as a brace to steady the frame. They extend the lamp 22 to 28 feet from the street corner. Mast arms can be used under any line of telephone or telegraph wires with twenty inches of space above. They will not freeze up in sleety or snowy weather. No step ladders or windlass are used.

Below will be found a diagram showing a popular method employed in stringing intersection lamps :

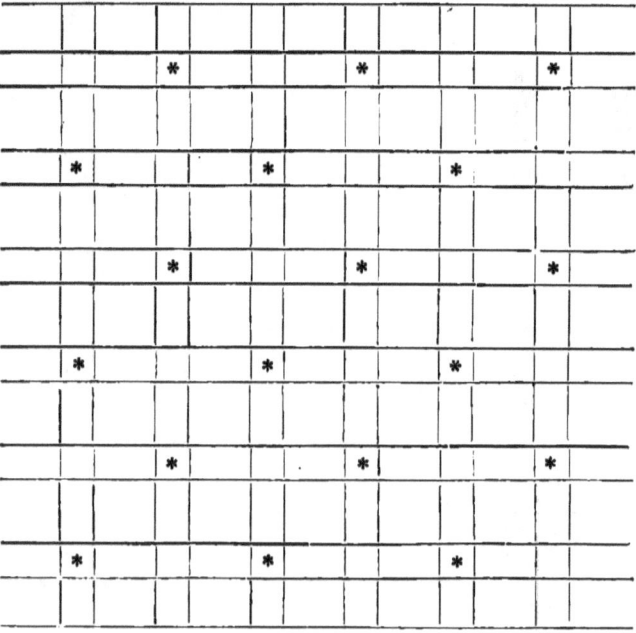

Lamps average 750 feet apart. At the corner where there is no lamp there is light sufficient.

Where lights are designed to be suspended over sidewalks a hanger, such as is here shown, will commend itself at once to

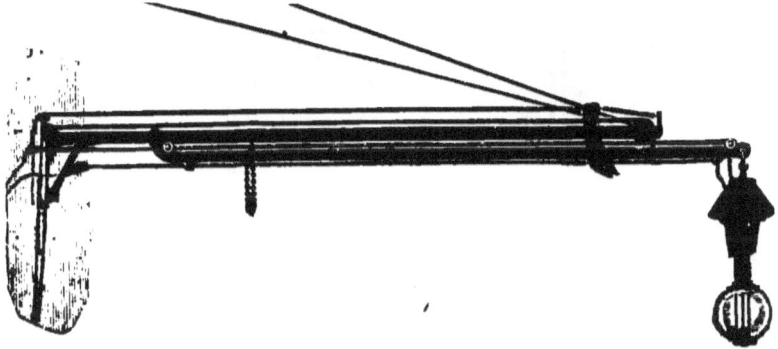

those who have had experience with fixed cranes, as it enables the trimmer to attend to the lamp without the use of a stepladder.

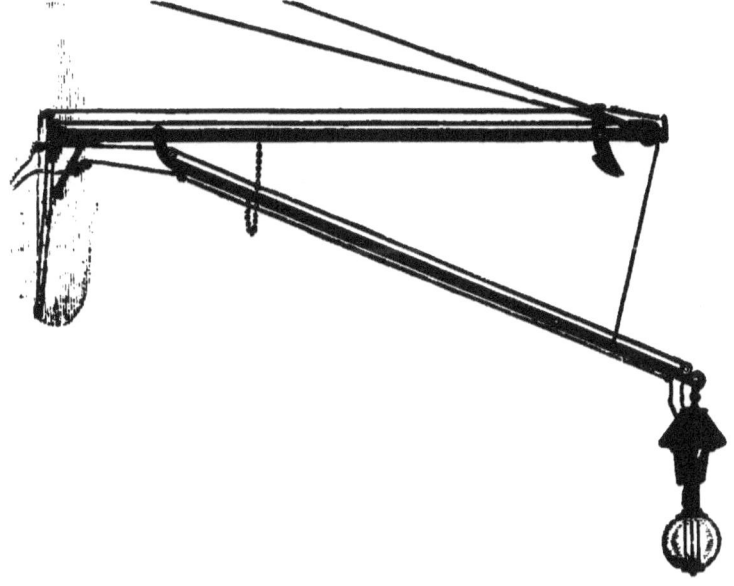

This convenience of access also insures a more prompt and satisfactory correction of any fault in the operation of the lamp when the step-ladder is apt to be at the other end of town.

A hanger does away with the objectionable loop, which is both unsightly and unsafe; takes the strain off the rope; tests the strength every time a lamp is lowered; the lamp swings to the inside of the walk for trimming; the lamp cannot drop. A short chain from the upper to the lower arm limits the distance the lamp will lower; a spring upon the catch makes its movements positive.

It consists of two rods or arms, the upper one fixed permanently to the building; the lower hinged to the former at a point near the building, carries the wires leading to the lamp on insulators. When the lamp is up, the catch effectually fastens the lower to the upper arm, from which position it can only be lowered by pulling upon the main rope first, and while this is held strained, tripping the catch by means of the second line.

For street service, where lights are suspended from poles, a form of this hanger possesses decided advantages over any means of lamp suspension. The light may be suspended at any required distance from the pole, and at any necessary height above the sidewalk or roadway.

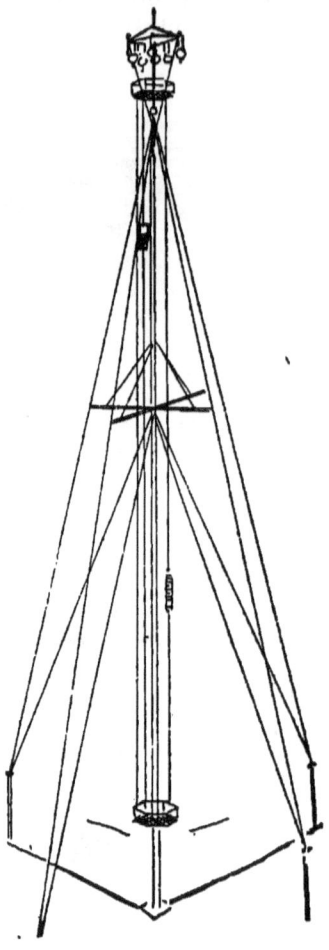

ADAMS TOWER.

TOWER LIGHTING.

The third method, placing lamps upon towers, is considered the best for lighting large areas with a comparatively small number of lights.

Detroit is the only large city in the world lighted wholly by the tower system. The city limits comprise about 21 square miles, the whole of which is thus lighted. There are 122 towers of 153 feet each.

Detroit has about 230,000 inhabitants, and has a dense business section of about one square mile. This section has about twenty towers, which average 1,000 to 1,200 feet apart; the belt immediately contiguous, embracing the closely-built and densely shaded residence section has its towers about 2,000 feet apart. Beyond this the spaces widen to 2,500 feet apart, and in the suburbs they are spaced about 2,500 to 3,000 feet apart.

The press of the country has uniformly conceded the city to be the best-lighted of any in the world. All its streets, yards, alleys, back-yards and grounds are illuminated as effectually as by the full moon at the zenith. The blending of light from the mass of towers serves to prevent dense shadows.

A comparison of the cost to the city of lighting by electricity with the former cost of lighting by gas, presents interesting features. The entire space formerly lighted by gas and naphtha, aggregated less than $7\frac{1}{2}$ square miles, with many streets inside that area not lighted at all. Now the area lighted is, as stated, 21 square miles.

With gas to the city on the basis of $1.25 per thousand feet, the last gas estimate for the seven and one-half square miles was $104,300. The amount paid the past year for lighting 21 square miles by the tower system was about $112,000.

Again, the relative expense of electric lighting by towers as compared with lighting a similar area by pole lights, may be seen. In 1885 the city called for bids on both bases of lighting—the exponents of each system submitting their own specifications of what would be required. The specification for pole lights indicated the necessity for poles not more than 400 feet apart on each street. This, in Detroit, would require about 2,600 for the city, now covered by the towers. The tower lights are burned from dusk to daylight every night in the year. The same service from 2,600 pole lights, at 35 cents per night, would amount to $910 per night, or $332,150 per year.

It would appear, therefore, that for a city, or town, proposing to do its own lighting, the tower system will enable it to cover the largest space with the least outlay, and that for a company proposing to do public lighting, the tower system enables it to bid for lighting at a figure much below any competitor seeking to cover the same space, either by electric pole-lights, or by gas and naphtha.

Lighting by the tower system is assuming great prominence, and merits the calm and thoughtful consideration of all interested in the subject of public lighting. It is the only system thus far presented which affords a thoroughly practical illumination of all spaces at a figure below the cost of gas ·or naphtha, as the latter is usually employed for quite inadequately lighting the streets alone.

Towers have been in use eight years in various parts of the United States for electric lighting. Anything that will lessen the expense of operating an electric light plant commands the attention of all interested in the subject. It is claimed for the tower system of lighting, that a city contract can be undertaken with less money at the outset and operated afterwards with less expense than can be done by any other method.

The construction of towers has been subject to gradual change and development. At the outset a square pyramidal tower was most used, made with inclined posts at the corners united by horizontal struts and braced both in the plane of the sides and horizontally. These horizontal braces prevented the ready arrangement and operation of an interior elevator, so resort was had to a triangular tower similarly braced on its exterior faces, but requiring no interior bracing. This left free space for an elevator, but, in order to insure strength, the tower was tapered from the top to the base, so that, with an altitude of 150 feet, the spread at the base was about 28 feet. This formed no objection where the tower could be located in a park or unoccupied public square, but where they had to be located at street corners, it was found necessary to have them span the street or side-walk, so that one corner of the

DISTRIBUTION OF LIGHT.

tower might rest near the corner of the street, and the two other corners rest adjacent to the buildings, and thus be in the way.

The development and improvement of pyramidal and other forms of tower gradually led to those used in Detroit. These are essentially triangular prisms, six feet on a side, built in vertical sections, each eight and one-half feet long, supported on a single base column and guyed at two points. The center lengths of all sections are equal, and members to a considerable degree interchangeable. They are located generally just inside the angle of the curb at a street corner.

The foundation is three feet square at top and base, and six feet in depth below the top of the cap-stone; is made of brick set in a mortar of hydraulic cement and coarse sand. At the base is an oak grill of two-inch plank laid cross-wise; the masonry is surmounted by a cap-stone three feet by three feet by six inches. Six bolts of one and one-eighth inch iron are headed beneath a large iron washer below the grill, pass up through the masonry and cap-stone and through the malleable iron casting secured to the base of the pillar.

The guy posts are of oak or other hard wood, sawed square and butted: are fourteen inches square and fifteen feet long, dressed above the ground,

"STAR" IRON TOWER.

the corners trimmed off and terminated at the top in a blunt, octagonal point. They are set vertically six feet in and nine feet out of the ground, at a distance about the length of the tower from its base, and usually just inside the curb line. Each has two eye-bolts about one foot below the apex for reception of two guy cables. The guy posts are tinned for about three feet, to prevent gnawing by horses.

The tower is made of lap-weld tubing and wrought iron brace rods. All castings are of malleable iron. At the base is a single pillar of seven-inch wrought iron tubing, fourteen feet high. This supports the shaft of the tower. The latter is composed of three corner elements, in vertical sections of eight and one-half feet, united by horizontal girts and strongly braced by diagonal brace rods. The tower is triangular, in cross section, and of the same dimensions from top to base—*i. e.*, six feet on each face. At the base the corner elements are of two-inch pipe, the six upper sections being of one and one-half inch pipe. At the top is a grating platform and iron railing. The towers have interior elevators, whereby the attendant may ascend to the top to care for his lamps. The elevator—of iron—forms a link in a continuous

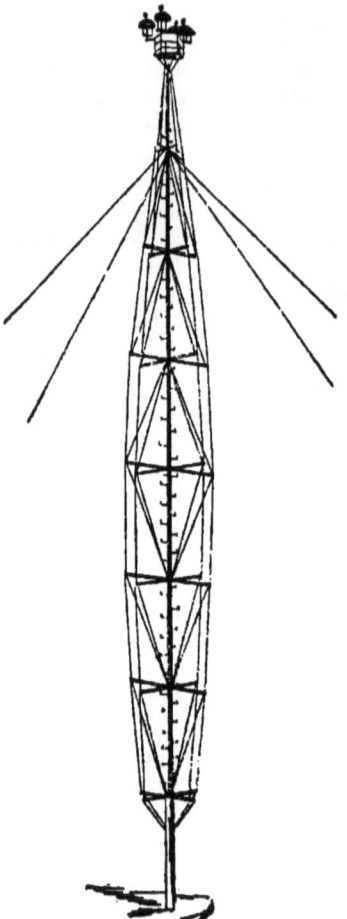

INDIANAPOLIS TOWER.

DISTRIBUTION OF LIGHT.

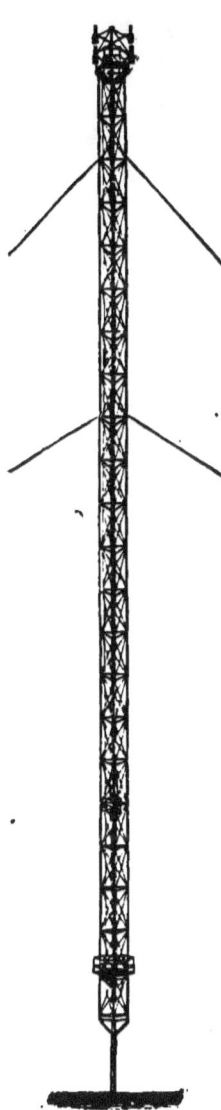

DETROIT TOWER.

wire cable, passing over a sheave wheel at top and bottom, and a heavy weight is connected into the cable, in proper position, to nearly counterpoise the elevator and its load. The towers have two sets of four one-half inch galvanized wire cable guys. One set leads from a point sixteen feet below the upper platform. The other from a point a few sections lower.

The towers are erected by first putting together the top section, then hoisting up and building on the next beneath it and so on until completed. The weight of a complete tower, elevator, etc., including guy ropes, is about 7,200 pounds. The entire wind surface, including lamps, hoods and mechanism, is calculated at eighty-three square feet, but this should probably be increased fifty per cent. for oblique and indirect exposure.

In the matter of the proper height, 150 feet is the most satisfactory. Increasing this height impairs the illumination near the foot and does not perceptibly increase the total lighted area, while diminishing this height diminishes the illuminated area and affords unnecessary brilliancy at the base.

The towers should, so far as practicable, be arranged in a triangular system. The distance apart in business sections may be 1,200 to 1,500 feet; in the best residence sections, such, for instance, as may be found at a distance of half to three quarters of a mile from the business centre, and a greater distance in the large cities, the towers may be 2,000 feet apart, and in the less densely populated sections and suburbs, they may be 2,500 to 3,000 feet apart.

Towns of three to six thousand inhabitants, occupying, say, a square mile of space, which determine to use towers, may be lighted in every quarter by seven towers, one at the center and six at the angles of a hexagon, the towers being 2,000 feet apart.

So, also, illumination may be had by five towers, there being one at the center and four at the angles of a square, the towers being 2,000 feet from the middle tower, and where greater economy is desirable there might be four towers, one at the center and one at each of the angles of a triangle and about 2,000 feet from the center tower.

Cities of from 6,000 to 20,000 would require from eight to fifteen towers; from 20,000 to 50,000 from fifteen to forty; from 50,000 to 100,000, from forty to one hundred; from 100,000 to 200,000, from one hundred to one hundred and fifty; from 200,000 up, from one hundred and fifty to three hundred.

In all cases it is recommended that the towers should have at least four lights of 2,000 candle-power each. The central towers might have six lights. More lights will improve the effect at a distance from the tower, while a less number of lights will scarcely afford illumination sufficient in the vicinity of the tower.

The question of the efficiency of tower lighting has been before many Boards of Aldermen, and will doubtless be brought before many more. As it is the one distinctive method opposed to all low lighting, information upon its merits or its demerits is valuable, therefore considerable space is given to official statements concerning it. The opinions following are from cities where towers are or have been in use:—

From the council committee of Flint, Mich:

"Our reasons for preferring the tower system, are:

"*First*—We are fully convinced that this is a system by which not only the streets, but the alleys, railroad crossings, depots, bridges, and even private grounds, are equally well lighted.

"*Second*—That after a tower is once located, the light therefrom will successfully illuminate an area of 2,000 feet each way, and it matters not how many streets are opened, or houses built within this district, the light covers the entire space.

"*Third*—We believe that the real estate lying in unlighted portions of the city, and looked upon as glooomy and undesirable, will be greatly enhanced in value when brought within the radius of this brilliant and far-reaching light.

"*Fourth*—We believe that in rendering our streets safe to all who traverse them at night; in largely preventing crime of every kind; in aiding us to attain that degree of peace and quiet which commends itself to every order-loving citizen, the benefits of this system of light can hardly be estimated.

"*Lastly*—We claim for it that it may be justly called the poor man's light, for, by reason of its penetrating and far-reaching rays, the suburbs of the city will be equally well lighted with the more central portions, and instead of the feeble flicker of the gasoline lamps, a clear and brilliant light will penetrate the most distant residence parts of the city."

From H. L. MACY, Auditor of Fargo, Dak:

"I believe the people are well satisfied with the tower system."

From the Fargo Argus:

"There is, there can be but one opinion in regard to it. The experiment is a complete and grand success."

From D. F. BARCLAY, ex-Mayor of Elgin, Ill.:

"Our people are well pleased with the towers."

From the Elgin News:

"The electric illumination of our city by the tower route is a success, as we had hoped it would be. It was a good success under the clear and tranquil sky. It was a better success still in the clouds and storm. The satisfaction is general."

From J. R. TREADWAY, City Clerk of Denver, Col:

"We have seven towers now in use, but we are getting rid of them as fast as the contracts expire. * * * We find the corner street incandescent lights preferable to towers."

From E. DELANY, JR., City Clerk of Fond du Lac, Wis:

"The towers have been in use here for the last five years. The people in the outlying districts, who have no other light, are well pleased with the tower system, although the towers are centrally located. If the city did not have the towers, it would probably not go to the expense of erecting them."

From THOMAS J. STEVENS, City Recorder, Ogden, Utah:

"The tower system of electric lighting was tried here several years ago, and discarded as being a complete failure. It is a bad system of street lighting, being impossible to erect a tower high enough to throw the light into the streets uniformly. One side of the street will be shaded by the buildings, which produces an intense darkness (when contrasted by the light on the opposite side), which is very disagreeable. Our city is now lighted by electric lamps placed in the center of the street * * * In the business portion the lamps are about 675 feet apart, while outside of the business blocks, they are two or three times that distance apart. Our present system gives good satisfaction."

From NEWTON FORD, City Clerk of Akron, Ohio:

"In 1880, we put up two towers, Brush system, which lighted the central portion of the city. They were owned, as were all the plant, by the city. In the spring of 1883, a local Brush company started here and a contract was made with them to run our plant for us in connection with their own. In 1885, we desired to extend our electric lights and advertised for bids. Two local companies—the Brush and Thomson-Houston—bid, and as the latter was only one-half the former, we accepted their bid and made the change. In a few months, after they had tried, in vain, to make our old Brush plant work, we discarded it, took down the masts, and now only use intersection lamps. Our experience is against the tower system and against the tower light. I would think it advisable for the city to own their own plant, as they would thereby secure cheaper and better service. Were the town level, and the streets straight and at right angles, with few shade trees, I am inclined to think that the tower system might be available."

From CHARLES E. DUSTIN, of Danbury, Conn:

"The electric light company at Danbury have now in use four towers. The general impression is that they are a grand success, and at a city meeting it was unanimously voted to continue them in use. Danbury is a hilly place, and the towers are so placed that the light is very greatly diffused. I do not believe that the introduction of double the number of lights would give such general satisfaction as those on the towers."

From F. A. BURKE, City Clerk of Council Bluffs, Iowa:

"The light is certainly an improvement on the gas lamps. When there is no moon at points about equi-distant from the towers the light is about the same that would be expected from a half moon. You can notice the removal of a brick, stone or plank on the sidewalk, as you walk along. Still there are people who complain that the light is not sufficiently bright."

From the Tipton, Iowa, Conservative:

"While not so bright in the immediate vicinity of the towers, on account of the increased height, the lamps give a much stronger light in the outskirts of town than was afforded by the lamps on the court house."

From JAMES H. FOSTER, City Auditor, of Evansville, Ind:

"* * * I have no hesitancy in saying that of the three systems in use, the arches are far the best and most satisfactory."

From the Evansville Courier:

"In the immediate vicinity of the towers the light is very brilliant, and the shadows formed are very sharply defined. As you go from the light toward a shadow, it appears to be dark beyond; but, on arriving at the shadow, the light is still perfect, though not so strong. The light produced is unquestionably the most perfect imitation of sunlight ever produced by human skill. The difference in the two systems, tower and arch, was plainly visible last night. As the light in its intensity and power, like that of the sun acting upon the unprotected eyes, is painful to a steady gaze, the advantage of an elevated system is at once seen. The arch light is somewhat modified by the use of shades, but the effect is produced at a reduced percentage of light; the result being an excessive illumination in the immediate vicinity of the lamps, and a less perfect illumination a short distance from them."

From ex-Mayor L. T. DICKASON, of Danville, Ill.:

"After a use for about four years of the tower and mast-arm system for lighting the streets of our city with the electric light, I can cordially recommend it for doing all that is claimed for it. The * * * tower system gives the most perfect satisfaction, and we all feel proud that our little city is one of the best lighted in the whole country. Public opinion is unanimous in its favor."

From a committee of the Bay City, Mich., council:

"We also find the system of mast-arms, projecting over the street from forty feet poles, a great improvement over the ordinary pole system. By suspending the light in the center of the street by means of the mast-arm, the light is clear from the foliage and all obstructions, and the streets can be lighted to much better advantage. We think that the tower system should be used in the outskirts of the city, where the buildings are not of such a height as to obstruct the light and prevent a free distribution; the mast-arms being probably more advantageous where the buildings are higher and more numerous. We are of the opinion, by actual observation, that Bay City should be lighted equally as well by the tower and mast-arm system, with from 15 to 25 per cent. less lamps, than by the present method [poles.]"

From Controller WM. BINDER, of Saginaw, Mich.:

"The light encountered heavy opposition in the beginning; are much satisfied now, particularly with the towers, where buildings are low, and mast-arms will give light about 300–400 feet radius. I should never use towers in thickly-settled parts, where streets are all built up on both sides with high buildings; even three-story buildings appear to destroy the effect of light sideways. Towers do well in wide streets, and will range 1,500–2,000 feet each side."

From E. C. HERR, Goshen, Ind.:

"Goshen is well satisfied with the towers."

From ROBERT KOEHLER, City Clerk of Rock Island, Ill.:

"Under our first contract, the city was lighted by eleven towers, of two lamps each. The service was not quite satisfactory, and in the new contract mast-arms are introduced."

From the Republican and Leader, La Crosse, Wis.:

"The general opinion is that there is nothing like it; that it does all that was claimed for it, and more, too, and that it is, in every sense, a success and a good investment for the city. Members of the Council, prominent tax-payers and business men, have expressed their satisfaction with the result. Many experiments have been tried to test the strength of the light, such as telling the time on a watch dial several blocks from a tower, or reading a newspaper, etc."

From the Decatur, Ill., Journal:

"The advantage of high over low light in penetrating behind buildings and trees will be recognized by all. Back yards and side alleys are illuminated as well as front lawns and prominent streets."

From H. A. BLUE, City Clerk of Macon, Ga.:

"The city is very well pleased with the towers."

From J. C. WHEELER, City Engineer of Macon:

"The towers give good satisfaction. We need more of them."

From W. A. KIRBY, of Jacksonville, Ill.:

"The towers are scattered over the city where it was supposed they were most needed. Our city is densely shaded, and not as well lighted as we wish. We hope to put in more intersection lights in the near future."

From T. F. HIGBY, City Clerk of Fairfield, Iowa:

"Trees obstruct the light somewhat in certain quarters, yet all parts seem to have sufficient light."

From GEORGE C. HODGES, secretary of the committee having charge of the lighting of Utica, N. Y.:

"The towers are used mainly in the outskirts and thinly settled districts. There they are a perfect success. In the heart of the city they are a failure."

From Hon. CHARLES F. MUHLER, Mayor of Fort Wayne, Ind.:

"Towers will do well enough for the suburbs, where houses are low and scattered, but are not the thing for the central part of a city. Towers about a quarter of a mile from the limits and half a mile apart, and low lights swung in the center of cross streets every other square and alternate, so that four lights are shining on the corner that has no lamp, makes the most perfect lighting."

INCANDESCENT STREET LIGHTING.

The principle of obtaining light by means of the "incandescence" or "glow" of an electric conductor enclosed in a glass receiver, from which the air is exhausted, has been well known for about half a century. The modern incandescent lamp or burner consists of a slender strip or "filament" of carbon attached to platinum wires and enclosed in a glass globe from which the air is exhausted and through which the platinum wires pass hermetically sealed.

Any electric conductor offers some resistance to the flow of the electric current, and resistance in a conductor through which a current of electricity is flowing produces heat. As heat above a certain temperature emits light, and as the carbon filament in the incandescent lamp is of high resistance, the passing of a certain current of electricity through it produces light.

The use of the incandescent lamp upon the streets of cities has not until recently come into very general use. In supplanting gas with electricity the idea has been to get so much more illumination than under the old method that the arc lamp was alone considered. It was an afterthought to furnish by the incandescent lamp about the same amount of or more illumination, and produce and supply it by neater and readier means. It was soon found to be cheaper, as well as better than gas, and there are now several companies which are devoting time and attention to incandescent street lighting.

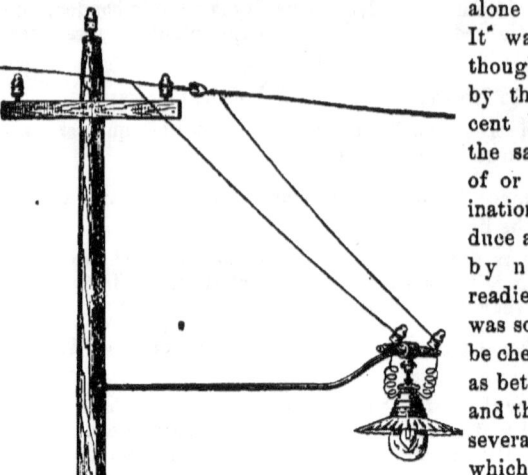

One obstacle to the first experimenters in the field was the large amount of wire necessary to send the current over circuits of any length. This has been obviated by the series system, by which the lamps are run on the same principle as the arc lamps.

In regard to the distribution of incandescent light for street illumination it is necessary to remember that the requirements of the various cities differ a great deal acording to their size, the ground plan of their streets and the character and pursuits of the population. In country places it may be sufficient to place one 20 or 30 candle power light at each intersection 300 to 400 feet apart.

In large cities it is necessary to have the alleys illuminated and also to place lights in the middle of the blocks. The methods of the gas companies in placing their lights can safely be copied as they have in most cases succeeded

STREET LAMP AND FIXTURE.

in making a perfect distribution of light so as to give an equal amount all over, and in placing the lamps where they are required. An improvement, then, cannot be sought in the method of distribution, but is simply a matter of furnishing more candle power. Lamps of 30 and 45 candle power, placed in or near the same places where the gas posts stand now would give any of the larger cities a most perfect night illumination, but if more illumination should be called for, incandescent lamps of from 100 to 500 candle power may be employed.

In some places where the streets are lighted with incandescent lamps the companies are connecting from 80 to 300 lamps in one circuit. They are also using double lamps with double holders. They have no groups but connect strictly in series, all lamps being on the main line and perfectly independent of each other. The current is of low amperage.

Underground Lines.

WITH the appearance of electricity into municipal lighting came the unsightly and dangerous conductors of the electrical current. These had not hitherto been a problem of public discussion, because in the rule of the gas regime there were entirely different mechanical methods employed to conduct the illuminant from the source of production to the source of consumption. The gas pipes were buried in the ground without any other than mechanical precautions, and they never obtruded themselves upon public notice except in cases of leakages, and then the sense of smell was outraged while the sense of sight was unaffected.

With the advent of electricity came the poles and wires—no doubt detrimental to the public safety and against the public good. The wires were trained overhead on poles and buildings; and, until they became so numerous as to imperil the safety of property, and some were demonstrated to be so deadly as to threaten life, they were suffered because they were the readiest means of providing and maintaining a plant. They were cheaper for the electrical companies, and they were satisfactory to the public corporations by reason of the lack of proper knowledge or the absence of fatal illustrations.

Some of the larger cities began some years ago to see the necessity of providing a different method of stringing the electrical conductors, and the underground circuit soon became a subject of discussion, but the technical difficulties in the way were so numerous, and the pace of invention in this direction was so slow, that very little actual progress has been made, and the subject to-day is engaging the best thoughts of the most skillful scientific workers. Such discussion as it has had has shown clearly that the undergrounding of lines is a feature of municipal lighting upon which reliable information is imperatively demanded.

ACTION SHOULD BE CONSERVATIVE.

So much that is in conflict has been written and said by those directly interested in the subject in a pecuniary and scientific way, that an official exposition of the precise situation of affairs is of value to every city contemplating a change from overhead to underground service. Especially is this information pertinent in view of the action of some municipal corporations in ordering all overhead wires within certain prescribed limits to be placed underground, which, no one familiar with the financial circumstances of the private companies, will contend can be accomplished without considerable expensive litigation. The fact that in the majority of the cities where such a change has been ordered, no steps have been taken by any of the telegraph or telephone companies, whose wires form the greatest danger to property, or by the high tension arc lighting companies, whose wires endanger life, to go underground with their lines in accordance with the city ordinances, makes this prophecy apparent.

There is also another view to be taken of this underground question. Should a city, whose council has declared that the wires must go underground, conclude that it is advisable to purchase and operate its own electric lighting plant, it must of course conform to its own ordinances and bury its own wires within the prescribed limits. To do otherwise would render the action of the authorities abortive and insincere. Therefore it is of the utmost importance that in putting its wires underground, a city should do so with the fullest knowledge of the subject; with a full comprehension of the difficulties in the way, and a thorough understanding of the best means of avoiding those difficulties.

If the city should conclude that it is not wise to invest in a plant, it is of more importance that its representatives should know what they are doing in ordering private corporations to bury their wires. If underground wires are practicable, this knowledge will strengthen the hands of the city and uphold the arm of the law; if not wholly practicable, this knowledge will serve to show wherein concessions and compromises can be made; if not practicable at all, the city will know that its present ordinances are unlawful, or at least unreasonable, and it can save itself the expense of useless litigation and the humiliation of defeat, by repealing them.

With the object of putting before municipal authorities, and to furnish all interested in any way in the training of electric light wires underground, the writer has prepared a somewhat exhaustive analysis of the situation, together with opinions pro and con from corporations, public and private, from manufacturers of underground wires, cables and conduits; from those who from choice or necessity use them, and from parties whose views are based upon their financial interest in, or their scientific knowledge of, the subject.

It is not necessary to detail the arguments in favor of the underground wire system, for the reason that, in the minds of the majority of people, the question is no longer one of advisability, but has passed that stage, and is now more particularly a question as to what method may be safely adopted for an underground system. The fact is admitted in all the large American cities and by the electrical companies, that the wires must ultimately be placed underground. A large number of the obstacles to the system have been done away with by experience, and the main objection of the companies has practically reduced itself now to one of expense and a perfect insulation.

Electric light conductors are very dangerous both to life and property whenever improperly insulated; and improper insulation is to be found almost everywhere they have been or are being used.

OVERHEAD SERVICE.

From its physical characteristics, arc lighting has gradually become used only for streets and large areas of enclosed space where it is best adapted for the diffusion of its concentrated beams of light. On the contrary the incandescent light, by the facility it affords for an equal distribution, is largely used indoors. The increasing use of these methods of illumination and their growing popularity has commanded attention to the question—what are the sources of danger of overhead wires?.

In many places electric light and power wires are carried

dangerously near buildings, awnings, telegraph or other poles, lamp-posts and other street obstructions. Again, in many cities several *distinct lines of poles* carrying electric conductors are to be found on the *same* side of the *same* street, and as these poles necessarily differ in height the wires upon them form a complete network, rendering the efficient use of the hooks and ladders and life-saving apparatus of the Fire Department almost impossible; whereas the placing of all the wires on the same side of any street upon one line of poles would in a great measure obviate this difficulty.

One source of complaint against overhead wires in numerous places is explained by the entire absence of any regulations or inspection, except that made by the electrical companies themselves. This is true also of all house connections. In fact, the city authorities in a great many cities exercise actually no control over the distribution of the most dangerous element necessity and invention has given us. The apprehension by the public of danger to life from the arc lighting system, with the current of great intensity necessarily used, and of danger from fire by the multitudinous wires of other electric installations, would be largely without basis if a uniform method was established for the inspection of the methods of wiring and insulation. Then, only the larger cities should force their electric light companies underground, and this force should only be applied after a practicable method has been tested and adopted. The smaller municipalities are abundantly able to protect their citizens and their property without forcing the companies to an expense that is uncalled for, except by a popular cry against electric wires.

The gas companies do not permit the flow of gas into a service newly introduced in a building until all leakages are detected and stopped. Under proper safeguards and restrictions the risk to persons and property from electric lighting can be reduced to a minimum, and in the absence of these lies the danger. The use of steam without proper precautions and by careless and incompetent persons, and even the ordinary illuminating gas and coal oils, may become sources of as great risk as electricity presents when the well-known electrical laws are not observed.

A thorough inspection, classification, and in some cases reconstruction, of the overhead service is a duty which councils should turn their attention to as quickly as to regulate the burying of wires. By so doing the dangers and complications of overhead wires will be materially diminished.

Of course it is not the duty of municipal corporations to solve electrical problems for the benefit of the electrical companies, and in judging of any system of underground or overhead connections proposed by such a company, the city is concerned with its electrical features only so far as they affect its repairs or changes, and the disturbance of the city streets which these may involve. That is to say, questions of detail concerning insulators, conductors, preventives or remedies for induction, etc., may be safely left to the companies themselves, their interest being to secure and maintain an efficient service in these respects; provided always that the system proposed be one which will permit experiment, repair, renewal and change without excavation. If the streets must be dug up every time an old cable or wire fails, or a new one is to be tried, then it is part of the duty of the city to know beforehand what sort of conductors, insulators, etc., will be used in any proposed system, and to approve only that system which promises in these respects the greatest efficiency and permanence. Otherwise, these considerations concern the electricians and managers of the companies only.

A SAMPLE SET OF RULES.

For the information of councils a set of rules is here submitted, which would render the overhead service much better in their localities. No doubt experience and a different order of things in each city will demonstrate the advisability of extensive modifications of and additions to these rules, but in many respects they embody a code which it would be desirable to adopt. The rules suggested are these:

1. No two lines of poles shall be on the same side of any street or avenue.
2. No two lines of poles bearing conductors or similar electrical service shall be on any street or avenue.
3. Electric light poles shall be of iron, at least twenty-five feet in height, with a diameter of not more than eight inches at the base, and having cross-arms of wood with glass, porcelain or rubber insulators, and painted a uniform color.

4. Poles for telegraph, telephone and other similar wires shall be at least sixty feet in height.

5. Poles shall be placed upon the sidewalk as near the curb as possible, and no pole shall be placed within ten feet of any lamp-post or other pole.

6. All wires shall be fastened upon poles or other fixtures with glass, porcelain or rubber insulators.

7. No wires shall be stretched within one foot of any pole without being attached to the same with glass, porcelain or rubber insulation.

8. No wires shall be stretched within twenty feet of the ground or within four feet of any building, except when attached thereto with glass, rubber or porcelain insulators.

9. No arc electric light or power wires shall be stretched over any part of any house or other building.

10. The companies or persons owning or controlling poles in any street or avenue shall allow the same to be used by other companies or persons operating conductors for similar electrical service, when authorized so to do, on tender of proper compensation, to be determined by agreement between the parties interested. In default of such agreement the amount of such compensation shall be determined by the council. This rule imports a contract on the part of each company or person owning or controlling the poles in any street or avenue, not only with the council, but also with each company or person who shall, under its terms, be qualified to demand the privileges it confers.

11. Any member of the council, or officer or inspector employed by it, as well as every member of the police force of the city, shall be entitled to examine permits under which work of any kind is being done.

12. All poles now standing, or to be hereafter erected, shall be branded or stamped with the initials of the company owning them, at a point not less than five or more than seven feet from the street surface. When an old pole is taken down it must be removed from the street the same day. New poles must not be brought upon any street more than two days in advance of their erection. Any pole that shall lie on any street more than two days shall be removed by the Department of Public Works, at the expense of the party owning it.

13. All electrical companies or persons having poles in the public streets shall give a bond to the city in a reasonable amount, to be determined in each case by the board, conditioned for the payment of the cost of removing dangerous and abandoned poles, and also for the payment of the expense of restoring the sidewalks and pavements where the same have been disturbed or injured in consequence of the erection or removal of any pole owned by them.

14. The violation of any of these rules and regulations shall operate *ipso facto* by a revocation of the permit held by the company or person guilty of such violation.

15. Whenever any company is permitted to erect posts or poles or other fixtures, bearing lamps or other devices for the purpose of lighting by electricity the streets, avenues, highways, parks or public places of the city, said permission shall be granted only subject to the following provisions, and the same is hereby expressly made a condition of said permits: "At any time when, by action of the city authorities, the contract for lighting any such street or other public place shall be given to another company, the company erecting said lighting fixtures or lamp-posts shall, on tender of the first cost thereof, yield possession and ownership of the same to the said other company obtaining the new contract."

UNDERGROUND SERVICE.

In this advanced age of electrical science it requires a constant outlook to keep pace with the latest and best improvements that are being invented for underground service. To serve the interests of municipal corporations this book will, therefore, endeavor to show the advance that has been made; what different systems of overhead and underground electrical service are in operation; what progress has been made in disposing of overhead wires, in what manner it is done, and to give the probable cost of placing the wires, either in conduits constructed especially for drawing in cables, or of burying the insulated wires in the ground, the kind of cable or conduit in use, and whether there is any particular benefit to be derived in having cables placed in conduits where they can be hauled in or out at pleasure, or buried in the earth. In nearly every city one will find different systems in use, with no particular uniformity of construction, either in shape, size, or the kind of material used. Some effort is being made to have the poles removed from the streets and the wires put out of sight, but, with the exception of the city of Chicago, the progress has been so slight that it is scarcely noticeable. There seems to be a fear on the part of some of the managers of telegraph, telephone and electric light companies that the time has not yet arrived to a certainty that all kinds of electrical currents can be worked any great distance underground a length of time that will warrant the cost of construction and removal of the poles and overhead wires. This has been the argument of electrical companies since the subject of underground lines was first discussed. So far as the telephone and telegraph companies are concerned, this fear is purely imaginary from an electrical point of view; with the high tension arc light system it is undoubtedly well founded.

IN CHICAGO.

There is more subterranean electrical service and a greater length of mileage in use in Chicago than in any other city, and the claim has been made by the electrical companies that the arc light industry has been practically strangled in consequence. The conduit system, in which various arc lighting wires have been placed, is made of a special form of asphaltum concrete, originally devised for sewer pipe, and according to the testimony of civil engineers who have carefully examined it, is practically indestructible for underground work, although it should be properly protected from the prolonged effect of extreme heat. This material is a good insulator, and it was this qualification which led to its availability for subterranean electrical work. This form of conduit is known as the Dorsett.

There are seventeen miles of Dorsett conduit in Chicago, containing 150 miles of wire. It is owned by the Sectional Underground Company, and is used by all the arc light companies, consolidated under the name of the Chicago Arc Light and Power Company.

The Western Union has ten miles of three-inch iron pipe, containing 150 miles of wire.

The Chicago Telephone Company has three miles of lead cable, and three and one-half miles of three-inch iron pipe, all encased in solid cement, and carrying 200 miles of wire.

The Bankers' and Merchants' Telegraph Company has fifteen miles of iron pipe and 400 miles of wire.

The Postal Telegraph Company has four and one-half miles of lead cable, buried in a square four-inch iron box, with 100 miles of wire.

The Baltimore and Ohio Telegraph Company has two miles of four and one-half inch iron pipe, surrounded by asphaltum, holding 50 miles of wire.

The corporation of Chicago has 6,604 feet of conduit, 19,296 feet of iron pipe, 1,200 feet of wooden trough, 73,731 feet of cable, 9,000 feet of three-inch sewer pipe, all containing 65 miles of single wire.

The cables used in the conduits are of different patents and various construction. Access to the cables is obtained by means of man-holes, placed at the intersections of the streets. The manholes are of the same material as the conduits, bottle-shaped, with cast-iron collar and double cast-iron covers. To provide for the distribution of wires a hand-hole has been devised, through which the cables are led. From this hand-hole smaller pipes are laid, similar in construction to the main pipe, but incased in iron to give the requisite strength.

Other conduits are laid, consisting of wrought-iron pipe about four inches in diameter, several of which are placed side by side. The cables running through these pipes are reached by brick man-holes, four feet long, three feet wide, and five feet deep, with cast-iron caps set flush with the street, with a drip to catch what water may work under the lid.

As to the operation of the arc light service in Chicago, City Electrician John P. Barrett writes:

"I have known electric lighting to be done in Chicago for about five years past, and am conversant with the means employed and which have been employed in Chicago for conducting electricity from the point where generated to the lighting point. Underground conductors for the conveyance of electric light currents are in use in Chicago quite generally. An ordinance passed by the Common Council prohibits the crossing of streets or alleys by aerial electric wires, and there are no electric light wires overhead. There are instances in this city where wires are extended from one building to another, where no streets or alleys intervene, over said building, but not over streets or alleys. There are many instances where an isolated plant furnishes power to lights in an individual building, or to more than one building in the same block, where no street or alley intervenes. Wires for such purposes are not underground, but in all other instances, and there are a great many of them in the city, the power is generated at a common center and distributed only by underground wires enclosed in cables and conducted through underground conduits to the lights. Where wires cross house-tops they are required to be at least seven feet above the roof, to prevent their interference with the fire department. Between 1,500 and 1,800 arc lights are supplied by means of wires laid underground.

"Electricity for lighting has been conducted underground in this city for at least four years. There are American, Western Electric, Sperry, Ball, Excelsior, Fort Wayne Jenney, Thomson-Houston and others, all arc light systems, in operation. In all cases where these operate plants that take in territory on both sides of the streets, they are operated through underground conductors. There have been such underground connections and conductors *in successful operation* any time within the past three years or longer, and these have been steadily increasing during that period of time.

"In my opinion *it is perfectly practicable to operate arc lights by means of underground wires*, and has been for the past three years. We have a variety of conduits through which electric light conductors are carried, principally the Dorsett system, which is composed of a nine-inch pipe, perforated with seven holes or ducts; these holes or ducts vary in size from one and one-half inches to two inches in diameter. This pipe is made of asphaltum and other ingredients, in sections three feet in length, laid beneath the surface and cemented together with a cement of the same material. These pipes are continued from street crossing to street crossing, and they are also connected with the corner of blocks on squares from man-holes. At street crossings man-holes are placed, through which the wires are drawn through the holes or ducts. I believe it is the universal practice to place signal wires, the positive placed in one duct and the negative in another. Each of these ducts can readily accommodate two lead-covered cables of the largest size.

"The construction of cables differs. We have what is known as the Western Electric, or Patterson cable, with a core of copper, cotton covered, paraffine insulation, surrounded by lead covering. We have also the Standard cable, the insulation of which I am not familiar with, also lead covered. We have also the Okonite cable, with core of copper. The size of these copper cores vary in proportion to the conductivity required, insulated with okonite sufficiently high to resist the pressure on the circuit. Then we have what is known as the Kerite cable, which is similar to the Okonite, with core the same, and is sometimes covered with lead, but usually not. We have also iron pipe laid beneath the surface, with man-holes at street intersections and wires led into houses underneath the sidewalks. We have also a limited quantity of wooden box, which is treated with some chemical process to preserve the wood, laid underneath the surface in the shape of conduits, through which cables are carried. Also sewer-pipe, laid in cement, with brick man-holes.

"Volts of pressure in the arc or high tension systems vary between 23 or 25 to 50 per lamp. Pressure tends to force the current through the insulation, hence, the higher the pressure the greater the need of insulation. I know of no difficulties in the way of operating electric light by means of underground wires. The only difficulty would be lack of conductivity, imperfect insulation, or imperfectly laid conduits, either of which would be purely mechanical.

"I should say the cost of Western Electric or Patterson cable, copper core, cotton covered, paraffine insulation, surrounded by lead covering, with conductivity sufficient for a fifty-light circuit, would be about $1,000 per mile; single wires, Kerite insulation, with same capacity and insulation, $1,000 per mile. I would suggest a two-inch iron pipe for conduits, which would accommodate three conductors, with twelve man-holes at average intervals. Such pipe, at present market quotation, would cost about $631 per mile, and man-holes about $45 each. Excavations, filling, repairing pavement, etc., would cost about $3,690 per mile. The aforesaid pipe will accommodate three conductors. The cost of increased space would be in proportion, as would also the cost of additional wire or cable. The prices I quote are based on the cost of material, labor, etc., in this market, and also on streets paved with wooden blocks.

"The city has had laid for it ten miles of cable, encased in iron pipe, underground, by contract. The work was let to the Western Electric Company, of this city, for $29,295. Embraced in this contract is supplying ten miles of iron-armored cable, protected in iron pipe where laid underground.

"In connection with the underground system, we have a number of man-holes at street intersections, which vary in size, the largest being about four feet square, and about three and a half feet high below the neck; the neck about two and a half feet long, reaching to the street level, and the whole being covered by an iron cover; through the area thus formed and of the dimensions thus given, all the wires and cable contained in both intersecting conduits pass, there being a conduit on each side of the intersecting streets. On the corner of Washington and La Salle streets is one of the largest sized man-holes, about the dimensions given, and through this man-hole pass eleven or twelve circuits, or eleven or twelve positive wires, and an equal number of negative wires, or twenty-two or twenty-four wires in all, and these wires are operated without any difficulty and without any perceptible escape of current. Several of these wires passing through the man-hole are used by the high tension electric light systems. *There is practically no difficulty in the arc and high-tension systems of lighting to wires carried in cables through underground conduits.*

"The Dorsett system of underground conduits was introduced in Chicago in 1883, and electric light wires *and high tension wires have been successfully operated ever since* through said conduit system."

B. E. SUNNY, president of the Chicago Arc Light and Power Co., which furnishes all the arc lighting in Chicago, writes to this effect:

"We are furnishing about 1,000 arc lights through underground wires in the business part of the city. There is something like 60 miles of cable used for this purpose. This company was organized after the passage of the prohibitive wire ordinance, and has never been allowed to put wires any place but underground.

"Our experience thus far has been of the most discouraging character. We have taken out and thrown away miles of wire covered with insulation of the character of rubber, and have substituted lead cables. One class of the latter cables have utterly failed, and we are now taking them out and replacing them with another style.

"*There is no question but what arc light wires can be successfully operated underground, but it is going to take another year, or possibly two, before it is found out what the conditions must be in order to get this result.* We think we have now found a lead cable that will serve us for a reasonable period, but after having spent our money on it we cannot say positively whether it will do so or not. In fact it is a thing that no man knows anything about."

IN PHILADELPHIA.

Numerous kinds of electric light, telegraph and telephone underground electrical conductors have been brought to the notice of the Philadelphia Electrical Department, with a view of introducing them in that city, and soliciting information as to the best method of bringing their particular kind or invention to the notice of the public, in some cases tendering to the department, free of cost, specimens for the purpose of having them tested. In several cases the proposition was accepted, the cables were buried in the ground, and placed in regular working circuits. Thus the electrical world has had an excellent opportunity to witness the results of a careful experiment.

In 1886, ground was broken for the introduction of the first permanent underground service for municipal purposes. The conduit, 5,200 feet in length, laid under the sidewalk at an average depth of one foot, and under street crossings of two feet, is of one inch spruce, with inside dimensions two by two and one-half inches, which was filled with pitch, after the cables had been laid in it. The cables used were those known as the "Waring." One, 5,200 feet in length devoted to telegraph and telephone purposes, is known as the six-wire anti-induction cable, from the fact of each conductor being insulated, separately covered with lead, and drawn together in corrugated form for convenience in finding the conductors. One of the corrugations has a sharp edge from which count can be made to the one required. The other cable is round, with eleven wires bunched and twisted together to form a conductor equal in conductivity to a No. 4 copper wire, and is used for electric lighting purposes.

"Both of these cables," writes D. R. Walker, chief of the department, "*have worked entirely satisfactory*, and have not been out of service since first put down. The current was first supplied by the United States Electric Light Company, by means of seven miles of overhead wires supplying Weston lamps. This service *has worked without failure*, the lights not having been extinguished, except through a stopping of the machinery at the Electric Light Company's works, or the breaking of the air-lines, and has demonstrated to me *the entire feasibility of underground service for electric lighting purposes.*"

"Although explosions have occasionally occurred in the conduit of private corporations," says Mr. Walker, "nothing of the kind has thus far taken place in those constructed by the city."

The largest private corporation in Philadelphia is the Penn Electric Company. For their conduits they have made use of creosoted wood, and in place of ducts have provided hangers on which to place the cables. The system is made use of in sections of the city where sub-ways would be too expensive, and is found to be excellent in many respects. It is the intention in Philadelphia to place all the city wires underground, yet it is a fact, known to everybody who visits the Quaker City, that at the present time there is but a small per cent. of the wires out of sight.

Chief WALKER also sends the following explicit information:

"In reply to your request for information about the working of high tension electrical currents (Brush and Thomson-Houston) underground in this city, and whether the experiments made by me have been successful or otherwise, I will say, we have seven miles of electric light cable in use, two miles of which were laid in May, 1886, and five miles early in 1887.

They are working entirely satisfactory to this Bureau, in fact so much so that I propose laying about five or six miles additional this spring. One of the cables is five and one-half miles in length, and the other one and one-half miles. To connect the longest length with the electric light station about five miles of air line is used, making the total length of conductor about ten miles, carrying a current of 2,700 volts, generated by a Thomson-Houston dynamo to fifty-eight 2,000 candle-power lamps of the Brush pattern. The Brush Electric Light Co. have also laid one mile of cable in the same conduit with that owned by the City, and is used in connection with it. *They never yet*, to my knowledge, *had a ground or trouble on it.* Several times during severe lightning and thunder storms, the Electric Lighting Company supplying the longest cable with current, have been compelled to shut down the dynamo, the shocks received were so great as to burn out the armatures of the machines. On one occasion two were burned out in one night. You can readily see what a tremendous tension there was on the cable for the time. A more severe test could probably never have been made.

"We have had some few grounds on the cables, but they were in a great majority of the cases caused by the carelessness of the workingmen, or those engaged in trimming the lamps. These troubles occurred principally at the terminals where the cable was carried to the lamps, and were readily removed. The first two miles of electric light cable the city had put down was laid side by side with a six conductor telegraph cable in a wooden box two and one-half by five inches, filled with roofing pitch to keep out gas and moisture, and *I have never had a ground* on either cable laid in this form. Both *are working perfectly satisfactory* to me. The balance of the cables, electric light, telegraph and telephone, are laid in creosoted wooden boxes, divided into ducts two and one-half inches square. I prefer conduits so constructed that the cables can be drawn in or out, at will, as circumstances require. The man-holes should not be more than 500 feet apart, and made of hard brick and cast-iron cover, sufficiently strong to resist pressure from above.

"The question of supplying current for arc lighting through cables placed underground has, in my judgment, *been settled beyond a doubt as practical.* It is less liable to interruption, and the cost of maintenance is far less than in the overhead service, the first cost of construction being the main drawback to its general adoption."

M. RICHARDS MUCKLE, JR., Secretary and Manager of the Keystone Light and Power Co., of Philadelphia, writes:

"While we are using the 'Waring' cable for incandescent

lighting in connection with our station, the current which we are using on the same is a 1,000 volt alternating current, which is carried into the buildings to converters, which there induces a fifty volt current, which we use in the lamps. In connection with Strawbridge and Clothier's plant at Eighth and Market streets, we have put in seven circuits of underground 'Waring' cable, each carrying current for thirty 2,000 candle power arc lamps.

"D. R. Walker, chief of the Electrical Bureau of this city, has miles of the 'Waring' cable underground, carrying current for 2,000 C. P. arc lamps of the Brush, Thomson-Houston and United States Company's systems, and he has had them in use for three or four years. For the most extended use of underground cable for arc light circuits, we would refer you to Mr. Walker."

Mr. A. J. DeCamp, general manager of the Brush Electric Light Co., of Philadelphia, gives another view of the question. He writes:

"The Philadelphia Company, of which I am general manager, has one circuit of about eight miles, about one-half of which is underground and the balance aerial wire, and upon which they have fifty-four lights; twenty-eight being directly underground and twenty-six upon the aerial part of the circuit. It has been operated under these conditions for about one year. The cable is No. 4 Birmingham gauge wire, encased in lead and known as the 'Standard' or 'Waring' cable. The incidents of the last twelve months are: three faults in cable, requiring the erection of aerial wires between certain lamps; and three explosions in man-holes, caused by defective cable to the extent of causing spark and igniting gas which had accumulated. The same station is furnishing the city with seventy-seven public lights, upon which they, the city, claim last month that *nineteen lights were out, fifteen of which were supplied from the underground circuit.* The above is, in brief, the history of the underground work in this city, which I learn is being very generally quoted as being a perfect success. As up to this date the public have got the light for which they pay, I suppose such an opinion naturally follows. With these facts before you, I leave the matter of success to your own judgment.

"Whatever conclusion you may come to upon the merits of this particular experiment, I will say that the general construction of this circuit is not such as could be adopted for general service by a central station of any considerable size. My experience with this and other experiments convinces me that we must first find *a perfect insulation, then a perfect method of applying it* before we can hope for any *permanent* success. In my own judgment, *neither of these have as yet been secured.* Granting that they may be, there arises a no less difficult problem of applying them in a manner sufficiently flexible to meet the demands which we find a business community make upon us and all other public servitors. In a word, as experiences multiply I grow more sceptical upon the subject."

T. CARPENTER SMITH, Superintendent of the Allegheny County Light Co., of Pittsburgh, writes:

"We only have a small amount, some 4,000 feet, of Waring cable in use, but we are placing some ten to fifteen miles of the same underground now, as the first has given us very good satisfaction. The writer recommended this cable to this company very strongly on account of its complete success with the Keystone Light and Power Company of Philadelphia, with which he is connected. We are fully satisfied that so far at least as high tension alternating and low tension direct systems are concerned, this cable will be entirely satisfactory, but we would lay great stress upon the fact that *no underground system can be expected to work unless the utmost care be exercised in making the joints.* We have examined into a number of underground plants which were reported as failures, and found in every case that the work had been done in such a slip-shod manner that the wonder was not that the system had failed, but that it had ever operated at all.

"Mr. A. P. Wright, of the Springfield, Mass., Light, Heat and Power Co., is operating an underground system on substantially the same plan as the Keystone Co. of Philadelphia, and ourselves. I am not sure whether he is using the same cable or not, but I know that he has had no trouble, and that he lays great stress upon the making of the joints and connections.

"As to the cost of the cable, that of course varies with the size of the conductors used. We lay all of our cable here in conduits out of which it can be drawn to make repairs. These conduits we find cost us about $1.00 per running foot, exclusive of the man-holes, for a box of twenty ducts of one and a half inch diameter, made of prepared wood. The man-holes cost us from $50 to $60 each."

The STANDARD UNDERGROUND CABLE Co. of Pittsburgh, writes upon the subject:

" The city of Philadelphia has laid considerable quantities of our lead covered cables underground for electric lighting. They buy the cables, lay them at their own expense, and, of course, they are the property of the city. The current for the lamps is furnished by several of the electric light companies operating in that city, the United States Company, the Brush Company, etc.

"It has been amply demonstrated by practical experience with our cables that *it is entirely feasible to operate electric light circuits in underground cables.* As to the cost of laying electric cables underground, that will vary according to the paving of the streets, and kind of conductor used, and the number of cables placed in each conduit. The size of copper conductor usually required for light circuits is No. 3 or No. 4 B. & S. G. and a lead covered cable containing a single conductor of that size would cost twelve and one-half to fifteen cents per foot. A wooden conduit (large enough for one cable) from which the cables could be subsequently drawn or additional ones put in, will cost from twelve to fifteen cents per foot, and the cost of laying the conduit and repaving the street may vary from 45 to 75 cents per foot of trench.

"This company is prepared to furnish its cable, and guarantee its practical working for electric light purposes. The cost of drawing this cable into this conduit and making necessary joints and branches will run about eight cents per foot of cable."

IN NEW YORK.

In New York, in 1885, a commission was established by act of the Legislature to examine into the whole question of placing the electric light system in use in that city underground, and to determine as to the most desirable method to be adopted. The commission is called the Board of Commissioners of Electrical Sub-ways. This commission has investigated the subject very extensively, and has had an opportunity of testing the latest and most improved methods in use regarding underground systems.

After examining the different forms of conduits, and considering the numerous plans proposed, the conclusion arrived at was, briefly, to the effect that the problem of removing the electrical conductors from the surface of the streets and operating them underground was rather one of a mechanical than of an electrical nature. In other words, it was found by the Commission that the conduit to be built should be modelled with an eye to the existing engineering difficulties to be met in the streets of New York, filled, as they are in most instances, with many kinds of pipes, contact with some of which would be dangerous, and deleterious subterranean influences, such as gases, escaping steam, salt water, etc., which, if allowed to come in contact with the electrical conductors, would be of serious injury to them; and still further, to the convenient placing, repairing, connecting, distributing and removing of electrical conductors, rather than to the questions of retardation and induction which proper insulation of the conductors would practically obviate.

To follow the language of the Commission, "a conduit is nothing more, electrically and mechanically, than a protection for the wires within it, and a convenience for placing them underground."

MR. DANIEL L. GIBBENS, one of the commission, furnishes the following statement regarding the underground conduits now in use, and in process of construction:

" The material to be used in the construction of conduits was to be considered only as to its strength and durability as a protector for electric cables. *The practibility of operating electrical conductors underground is now settled.* The commission were at work constructing conduits, and the result of their work demonstrated that it was only a question of a few years when all the electric wires in the city of New York would be operated underground. The result of their experience is, that *electric wires can be operated as cheaply and successfully underground as overhead.* The sub-way in Sixth Avenue is completed for a distance of two miles, and consists of twenty-four ducts, and cost $60,000 a mile. This is a drawing-in system, with frequent man-holes, so that the wires are easily reached at frequent intervals, being best adapted to meet the requirements of the present electrical service."

In inaugurating the underground system, the New York Commission followed certain principles, which may be defined as follows:—

First—A conduit or subway for electrical conductors is nothing more than a mechanical protection for the wires within it, and a convenience for placing them and protecting them underground.

Second—Electric light and power conductors should, as a matter of precaution if not of necessity, be operated separately, and as far as possible from those for the transmission of currents of lesser intensity.

Third—The material and form of the subway should depend

largely upon the requirements of the locality and the service for which it is designed.

Fourth—Drawing-in-and-out conduits with convenient manholes are, in the main, the most desirable for the streets of the city, where a condition of the law allowing the companies ninety days to place their conductors in the subways after they are constructed, necessitates that the subways shall be easily accessible without serious disturbance to the pavement.

Fifth—The success of the underground service depends largely upon the proper insulation of the wires, and the largest liberty compatible with the preservation of the rights of others should be allowed to the companies making use of the subways.

Sixth—The nature of local connection depends to a great extent upon the service and locality for which they are designed, and here again liberty of choice under proper restrictions may reasonably be allowed.

Proceeding from these general principles the board has constructed subways in different localities largely differing in design, size and material. The work of the board has been done under the direction of its Chief Engineer, Henry S. Kearny, by a construction company known as the Consolidated Telegraph and Electrical Subway Company, of which Leonard F. Beckwith is Chief Engineer. The completed conduits are the property of the construction company, and the right to their use is leased by it to the various telegraph, telephone, and electric light companies. Chief Engineer BECKWITH writes as follows upon the subject:

"Several different systems of construction have been adopted in this city. The bulk of the conduits laid consist of lap-welded wrought iron pipe with screw joint couplings, laid in hydraulic cement concrete, the pipes previously treated to a coating of asphaltum; cost about fifteen cents per lineal foot delivered in New York, the usual size being two and one-half inches in diameter. Some iron pipe of the above description has also been laid in asphaltic concrete. We have also laid a section of the work with a cement pipe, similar to what has been used in water distribution, and consisting of a sheet of iron covering, lined with five-eighths inch thickness of pure cement; and costing about eleven cents delivered. This pipe is laid in hydraulic cement concrete in the same way as iron pipe.

"Another section has been laid with creosoted wooden tubes, costing about eight cents per lineal foot, consisting of squared logs four by four inches in section, eight feet long and bored out with a two and one-half inch hole. These tubes are boxed together by creosoted two-inch planking with tarred joints.

"There were nearly 267 miles of single ducts laid during 1887. The cost of excavating trenches and laying conduits is very high and variable in a city like New York, in which the ground under the street service is crowded with pipes and obstructions of different kinds and is not alike in any two streets. It forms no criterion or basis for the cost for other cities in which the cost should be much less.

"Access to the subways is obtained at intervals by manholes, which we endeavor in general, to build five feet square, and which frequently are irregular, owing to adjacent obstructions. All man-holes are built with eight inch walls of North River brick, cement concrete bottoms, and the cast-iron street frame with double cover. The inner cover rests upon a rubber tubing gasket to exclude street water. The inner cover is locked so as to protect the property in the subways from interference. A section of 'Dorsett' conduit, consisting of coal tar and sand blocks, was laid in New York in 1886, and is in use to-day, chiefly for telephone cables.

"Electric light conduits, and telephone and telegraph conduits are placed on opposite sides of the streets where they are both in the same street, in order to prevent interference by the currents as much as possible. The conduits laid down for electric light purposes so far have been iron pipe laid in hydraulic cement concrete. The number of ducts in the subway varies very much. In one street we have 104, in others 70, 50, 40, 20, six and even less.

"The distribution of the wires from the subways to the buildings is proposed to be done by ducts leading from the manholes to a basement, and through a building to the rear yards, where a pole affords means of conducting them to the rear windows of buildings. Another construction consists of leading the pipes from the man-holes to a recess or angle in the front of a building in which iron leader pipes conduct the cables to the roof, where, by some usual house top fixtures, the wires are distributed to the block.

"Another method consists in laying a distribution pipe in the subway trench over the conduit, which pipe is furnished with cast-iron hand holes at intervals, whence service pipes run into the cellars or basements of buildings adjacent.

"The Edison Electric Illuminating Co., for incandescent lighting, has laid an extensive system of their special tubing with junction and distributing boxes, one of the chief features of which consisting of wires and insulation filling the tubing, being laid at the same time as the latter and connected in twenty feet lengths as laid."

G. McFALL, Secretary of the Brush Electric Illuminating Co., of New York, writes:

"We are not furnishing any arc lights for the city or otherwise underground, nor are there any underground wires in successful operation in this city, excepting a few of the Edison Company for short distances. There have been some conduits laid composed of iron pipe, with the intention of placing arc wires in them, but up to the present time *no system has been able to operate with any show of success underground.* I do not think it feasible to run high tension currents underground at present until some method is devised by which a conduit can be constructed that will withstand the elements of the earth, and at the same time be of an insulating material. I have had some correspondence with various cities through the United States and Canada, as well as in Europe, but to all communications comes the same reply—*the underground business for high potential currents is a failure.*"

IN WASHINGTON.

In Washington, where the government authorities are forcing the electrical companies to place their wires underground, no inclusive system has been suggested, but each company is supposed to adopt such a plan as may best suit its requirements. The plan adopted for arc lighting wires consists of a wooden trough in which insulated conductors are laid side by side, the wires being separated from each other by the use of bituminous bridges, which are placed about eighteen inches apart. The trough is then completely filled with bitumen, making a solid insulating mass impervious to dampness. The distribution wires are branched out in a similar manner. With a small number of main lines this operation is comparatively simple. In Washington, however, both sides agree that the experiments have not been successful.

A. M. RENSHAW, general manager of the United States Electric Lighting Company, of Washington, writes the following positive letter:

"We have laid in this city during the last few years some twelve miles of the Callender cables in solid asphalt conduits, but we regret to report that *it has proved to be a failure* and *it has all given out,* and *is now abandoned.*

"We have lost in the last few years over $20,000 in attempting to make various underground cables work. We have succeeded in making them answer for several months, (and in one case eighteen months) but they gradually give out and must be changed or abandoned. Parties interested in cables will tell you that it is perfectly practicable to make them work, but you can rely upon it that *there is no cable in existence that will not give out* when steadily subjected to an arc light current of 2,000 volts.

"We have brought experts here from Europe and spared no expense to accomplish a successful underground system; but there is no place in the *world* that has ever maintained an arc light current of 2,000 volts for a period of two years. Better expend twice the amount on building some overhead conduit for lines, than to put them in the damp ground where they will always have trouble."

CHARLES W. RAYMOND, major of engineers of the District of Columbia Commissioners, writes:

"The underground arc light system in Washington has not been a success, whether on account of defects in the conduit or for other reasons, we are not at present prepared to say."

IN BROOKLYN.

In Brooklyn, as in New York, an electrical subway commission has been established to investigate the subject, and determine, if possible, as to the best methods to be made use of for conducting wires underground. The president of the commission, Prof. George W. Plympton, has visited the principal cities of this country and Europe, and submitted a report last year, giving an account of the progress made both in this country and in Europe towards establishing underground systems. As this report furnishes the latest information concerning the underground systems in foreign cities, it is presented here as a matter of interest, although it may not be found of any special value, owing to the fact that the electric service in foreign cities, from the meagerness of its extent, offers no comparison to that of the leading cities of our own country. The report follows:

At a meeting of our Board, held in July last, it was deemed desirable that a personal examination by some member of this Board should be made of the telegraph and telephone systems of European cities.

In accordance with the terms of a resolution passed on the 8th day of July, I left New York for Liverpool on the 21st of that month, arriving in Liverpool on the morning of the 29th, and reaching London on the evening of the same day.

Some letters of introduction to Mr. Preece, the electrician of the British Postal Telegraphic service, secured for me the cordial and polite attention of the chief himself, and the very efficient aid of the officials in charge of special departments of the service. No extensions or repairs of the lines were in progress at the date of my call, but I was invited to examine the construction of the conduit pipes and the cables they contained, so far as they were accessible at the "working boxes," or "manholes" as they would be called in America. Accompanied by Mr. Fleetwood, an electrician of the postal service, I examined several of these "boxes," which had been opened by his order for the purpose.

The covers are fitted very tight, and are flush with the pavement of the sidewalk. In two instances the workmen failed in their attempts to pry them off. The "boxes" are shallow, scarcely two feet deep, and only sixteen by twenty-four inches in lateral dimensions. The conduit pipes are of cast-iron, and mostly of three inches in diameter. Into these pipes the insulated wires, in a loose bundle, are drawn; gutta percha is the only insulating material used. The wires are for telegraphic service almost exclusively; only a few telephone wires, and these for the use of the post-office, are under the pavements. Telephonic communication in London is furnished by the system of the United Telephone Company, whose wires are all supported overhead by house-top fixtures. The right to erect the fixtures is purchased of the owner of the property.

Experience with underground telegraph lines is not new in England. Before 1845 such lines were placed underground, but they soon failed, and pole lines were substituted. Again, in 1853, the Magnetic Company of England laid wires for telegraphic purposes in creosoted wood, buried two feet in the earth. The insulation gradually failed, and again the system was replaced by one of poles.

A section of the conduit used at this time was shown me by Mr. Preece. It was a trough-shaped block, about four inches wide, and perfectly sound, although it had lain buried at shallow depth more than thirty years. Iron pipes are now preferred, solely because of the protection they afford against the blows of pick-axes and shovels of careless workmen—a protection which seemingly might have been secured by a heavier conduit of timber. Mr. Preece said he knew of no case of decay of well-creosoted timber.

From London I went direct to Brussels, where I arrived on the 2d of August. To Messrs. James F. Meech and George Cutter, both of the International Electric Company, I am indebted for much valuable aid in getting desired information. These gentlemen are engaged in superintending the erection of the Thomson-Houston system of arc lighting. Their operations during the present year have extended as far east as Russia, north to Sweden and Scotland, and south to Italy.

The acting superintendent of the telephone service in Brussels is Mr. E. Bartelyou, who kindly gave me the details of the system. The wire used is a phosphor-bronze wire, of 1 4-10 millimeter diameter (or about No. 17 gauge). Its strength admits of long distances between the supports, which are all on the house-tops (no wires of any kind are underground in Brussels). In a few places there was a clear space of 270 meters, or 885 feet between the supports. No stretching or breaking had occurred in these long stretches, although the wires had been in place for three years.

The managing director of this company is Mr. DeGroot, whose principal office is in Antwerp. As I had brought letters of introduction from personal friends of Mr. DeGroot in New York, and as the franchises of his company include the telephone systems of Brussels, Antwerp, Stockholm, Milan and Turin, I went on to Antwerp without delay. All my inquiries regarding electric systems in the above cities were kindly answered. All wires of all kinds in the above-mentioned cities, whether for telegraph, telephone or electric lighting, are above ground. In the northern cities they are on the house-tops, and in Italy partly on the house-tops and partly upon brackets along the sides of the buildings. Two exceptions to this general rule were mentioned. One was a short circuit for electric lighting on the Edison (incandescent) system buried in Turin, and the other was a new line of telephone wires supported on poles in one of the streets of Antwerp.

Rights to erect fixtures on house-tops are purchased, as in England. The price in Belgian cities is about 50 centimes (10 cents) per wire per annum. The more recent supports for wires are of wrought iron; and are of somewhat ornamental design. The overhead wires are thickly clustered in some portions of Antwerp and Brussels, but being of smaller diameter, are not so unsightly as the same number would appear in American cities.

I went on into Germany on the 6th of August, making brief stops at Cologne, Wiesbaden, Frankfort and Heidelberg. Telephones are not much used in these cities, only 500 in Cologne, 464 in Wiesbaden, 430 in Frankfort, and less than 100 in Heidelberg. All the wires for these systems are supported overhead, on house-top fixtures of rather rude patterns. The military telegraph is underground. All others are in the air. In Germany, as in England, the buried systems were tried early. The first ones failed at once. Upon the discovery of the insulating qualities of gutta-percha by Dr. Werner Siemens, in 1846, new experiments were tried with underground systems. A line of four or five miles in length was laid between Berlin and Gros Beeren in 1847. As this appeared to be successful, the experiments were continued until more than 3,000 miles of gutta-percha covered wire had been buried in the ground. This worked well for a few years, but finally failed entirely, and a line on poles was substituted for it.

The telegraph wires now buried in Germany are in bundles forming cables, known as Siemens' cable, each wire being separately insulated, and the whole protected by a spirally-wound heavy wire, which shields the conductors from ordinary injuries. It is an expensive cable, and is better adapted for the military service, for which it is chiefly employed, than for the wants of a commercial community. That its use is thus restricted in Germany is pretty well indicated by the fact that wires on poles stretch along every mile of railway throughout the empire.

I arrived in Berlin on the 9th of August. As telegraph, telephone and two kinds of electric lighting systems are used in the city, I sought information at several points. The telephone used here is a German invention, differing somewhat from the familiar Bell pattern, and declared to be unsatisfactory by those who were accustomed to the latter.

Both the telegraph and telephone systems belong to the government. The telephone wires are carried on house-tops upon light iron fixtures, much like those in Belgium. The house-owner is paid an assessed compensation for the occupation of his premises, but he is obliged to submit to it.

An electric lighting company have buried cables along several streets of the city. The engineer of the company gave me the following list of rules imposed by the Imperial Government upon their work in the streets:

The electric light cable must be enclosed in an iron pipe whenever it is buried within one meter of and parallel to a telegraph cable.

In case of the crossing of the two systems, the electric light cable must be at least 45 centimeters (17.7 inches) distant and above the telegraph; and, moreover, must be covered with an iron pipe for at least two meters each way from the point of nearest approach.

Whenever an electric light conductor is laid in a street where there is no buried telegraph, such conductor is to be placed under the sidewalk, but at least two meters from the house line.

I visited the Siemens' establishment, but was unable to meet Dr. Siemens, to whom I had a note of introduction. He was out of town. The manager of the works in the city kindly offered to answer my inquiries, and to act as interpreter in an interview with their engineer. In course of the conversation, on two separate occasions, I was told that although they knew of the questions arising in America, regarding the safe limit of proximity of wires for different kinds of electric service, they were not prepared to express an opinion on the subject. They knew of no experiments with the quadruplex telegraph, and the high tension alternating currents for arc lights were not regarded with much favor, and were not in use in Berlin.

It was understood that the government telegraph officers had been trying some experiments bearing upon the solution of these questions, but with what result was not known.

From Berlin I went south, arriving at Munich on the 13th of August. An official at the telegraph station connected with the post-office informed me that all the lines were on the house-tops. The number of telephone subscribers was 730. The Morse system of telegraphy was used and the German telephone.

In Lucerne, my next resting place, the telephone is but sparingly used. Wires are upon light fixtures on house-tops.

In Milan, where I arrived late on the 16th of August, I was fortunate in meeting Mr. Strugnell, an electrician of the International Electric Lighting Company, who at once interested himself in the object of my visit, and rendered important service. Milan has about 800 telephone subscribers. All the wires are above ground, a large proportion of them being under the eaves, attached to bracket fixtures. The reason for this departure from the common European practice is found in the triple system of tiling on the roofs.

The Edison, the Siemens and the Thomson-Houston systems of electric lighting are all in use, with overhead wires only, and seem likely to extend rapidly. Some complaints had just been made against the Thomson-Houston system, on account of induction effects due to its high tension current.

An effective system of street lighting is in practice here. Arc lights of good intensity are suspended over the middle of the street from high brackets attached to the buildings on both sides.

I visited Venice and Turin, but found no features essentially different from those at Milan. The systems of all kinds were only of less extent.

I reached Paris on the 21st of August. Having letters to M. Berthon, the well-known electrician and inventor, I went directly to the offices of the telephone company in Rue Caumartin. I was cordially received, and invited to ask as many questions as I liked. In response to my inquiries M. Berthon gave me, in substance, the following information:

There were in Paris at that date about 4,400 telephone subscribers, and about 2,200 in all the rest of France. In Paris telephone wires are made up into small cables and conducted through the sewers, being held in place by small hooks, the shanks of which are driven into the masonry joints. If the sewer ends before the destination of the telephone cable is reached, the cable is immediately run up to the house-top and conveyed overhead, as in other European cities. In other French cities the wires are all above ground, except in Bordeaux, where 400 subscribers are served through a system drawn through iron pipes underground, but at the date of my visit the system was not working very well. Two hundred of the Paris subscribers are reached by wires above the houses, because there are no large sewers in which to suspend the wires.

No electric lighting wires are permitted in the sewers. Indeed, there are but few electric street lights in Paris. There are less than there were five years ago, and what there are, are all overhead.

In both London and Paris all arc lights are within short distances of the current generator, which is generally a dynamo, driven by a gas-engine.

M. Berthon expected his company would undertake the telephone service for the principal cities of Spain. He proposed to exact as a condition the right to carry the conductors upon the house-tops, employing a metallic telephone circuit, and cables of a small number of wires in each.

In traversing the sewer under the Avenue de l'Opera I found two groups of telephone conductors, of twelve cables in each group, fastened securely to the arched top of the sewer, while brackets on the sides supported a pipe carrying telegraph cables, a gas-pipe of six inches and a water-pipe of sixteen inches diameter. When the telegraph cables reach the end of a sewer they are continued on underground in three-inch iron pipe.

Leaving Paris on the 25th of August, and spending two days in London to make some further observations upon lines in the suburbs, I continued on to Newcastle. This city enjoys the reputation of having buried almost its entire system of telegraph and telephone wires. The telephone service is not very extensive, as the number of subscribers is not quite 600, and is increasing but slowly, some 25 to 30 between January and August. Nearly all the subscribers, moreover, are within a circuit of a half-mile radius.

Through the kindness of Mr. Heavyside I was permitted to examine, in company with an inspector, the central office and the entire system of distribution.

The arrangement of conduits and street "boxes" (corresponding to our man-holes) is much like that in London. The boxes are 10 inches by 26, the pipes entering at the narrow end. Pipes of 2, 3 and 4 inches in diameter are employed. When the number of wires in any pipe is to be increased, the entire contents of the pipe are drawn out and the new lot drawn in. The wire used is No. 18 copper wire, with a wound covering, whose outside dimensions are those of No. 7 gauge.

The wires are bound together in sets of four, the pair which form the circuit being twisted together with much care, to avoid induction from the neighboring pair.

As prominent electricians have expressed the opinion that, when a metallic telephone circuit is employed, the limit of distance of telephone communication underground is about 10 or 12 miles, I was quite desirous of testing the question. The privilege was cheerfully granted. A line was connected for my use, that extended through $12\frac{1}{4}$ miles of underground wire, and continued thence over a pole line of 38 miles further; an entire length of $50\frac{1}{4}$ miles. The result was satisfactory in every way. During a conversation of several minutes there was no call for a repetition of a word.

The telephone service of Newcastle is certainly good, but to enlarge it to meet the requirements of an American city would demand extensive modifications; not only the greater extent, but the frequent changes of our telephone service would completely transform the problem.

Time did not permit a stop at Manchester. I had learned that some preliminary steps had been taken to bury the wires in that city, but it was evident, from such inspection as I was able to make, that the aerial system is still quite extensive. There has been as yet, I believe, no burial of wires in Liverpool.

My journey in Europe, of thirty-four days duration and four thousand miles in extent, ended on the 1st of September at Liverpool.

As a summary of the observations made during this short tour the following are offered as conclusions fairly drawn from the information obtained:

The problem of construction of a telephone system on such a scale as exists in American cities with underground wires has not yet been solved abroad. In Europe no attempt has been made to solve it, because as yet there is no such general use of the telephone. The number of telephones in the United Kingdom of Great Britain, reported on January last, was 13,000, while in the United States at the same date there were 163,500. There are more telephones in use in New York City and Brooklyn than in all Great Britain; more in New York City alone than in all France.

For the accomplishment of the ultimate object of converting our overhead electric systems of all kinds to the underground systems, without impairing the efficiency of their service, without permanent injury to our streets, and without increasing the cost of the service, so as to prevent the use of the same, we must depend upon the experience gained and experiments made on our own side of the Atlantic.

Prof. Plympton writes from the Brooklyn Polytechnic Institute:

"I have been expecting some definite information in regard to the underground system in Philadelphia. I received a communication last night from the expected source, but it was not sufficiently positive in its character to allow me to base any strong hopes upon it of a speedy solution of the arc lighting problem.

"A system of arc lighting through Waring cables (or lead-cased conductors) is under trial in Philadelphia. Accounts of the results obtained are conflicting. Nothing has been done yet in this way in either New York or Brooklyn.

"Regarding underground conductors for arc lights I can only say that the question of its practicability is, in my opinion, *not yet settled*. We decided long ago that we would not allow arc light wires in the same conduit with telephone wires, nor would we allow them to come into the same man-hole for purposes of distribution. No arc light wires are yet underground in Brooklyn. From the best information we can obtain it would seem that to force the arc lights underground upon any plan yet suggested would subject the plant to an even chance of destruction within two or three years. Still, I think the problem *will eventually be solved.*"

The last annual report of the Brooklyn Board contains considerable discussion of the underground question. The Board says:

"The experience of the year has confirmed the previous opinion of the Board as to the material and plan of construction of subways heretofore approved by it for telegraph and telephone conductors. The systems introduced in other cities do not seem to present any points of superiority, when all conditions are taken into account, and in several respects are decidedly inferior to the system adopted by this Board. Practice has shown, however, that the conductors first laid in the Brooklyn subways require improvement. They have been found inadequate, for electrical reasons, beyond a distance of about 7,000 feet.

"The conduit approved by this Board permits the removal or renewal of conductors without excavation, and if the facts now developed by experience shall require the substitution of new conductors in any part of the subway system, the ease with which this can be done justifies the rejection by this Board, in the beginning, of all 'solid conduits' not permitting such changes.

"The precise change of conductors now contemplated is the use in future (and probably, to some extent, the substitution in present underground lines) of a heavier copper wire, containing about twice as much metal, and requiring about twice the thickness of insulating material. It is not improbable that the use of complete metallic circuits (instead of 'grounded circuits,' employing the earth for the return current) will ultimately become necessary. These contingencies were apparent to the Board when it approved the present underground system; and the electrical companies were fully advised of their possible occurrence. In view, however, of the increased expense to the companies involved in the use of twice or even four times the quantity of metal and of insulating material required per subscriber, and the possible success of the cheaper grounded circuits and lighter wires, the Board simply notified the companies at that time that the underground system could not be pronounced impracticable by reason of any defects which the above-mentioned improvements would remove, and that they would have to adopt these improvements, if necessary to the operation of a subway system. The introduction of heavier wires, and even of metallic circuit, will not affect the existing conduits, except by reducing their capacity, as measured in the length of conductor contained, and the number of subscribers served.

"Among the questions which need further study, that of electric light conductors is one of the most important. After the organization of the Board, the electric lighting companies declared that they had no plan to propose for putting such wires underground. This made it the duty of the Board to devise such a plan, if possible. The subject has been studied with care, but not as yet with the necessary completeness.

"With regard to the electric light conductors, the Board has found no device which would with certainty, in its opinion, enable the wires carrying arc-light currents to be safely and successfully operated in the same conduit with telephone and telegraph conductors without disturbance or injury of the latter. Hence the consideration of the arc-light conductors must have reference to independent conduits kept at a distance from the telegraph and telephone conduits. The following points are suggested, and their number might be increased :

"1. It is a question to what extent the arc-light currents, especially those of certain systems, involving very high potentials, or so-called alternating currents, can be successfully maintained in insulated conductors underground. As will be seen in the report of the President of the Board, herewith transmitted, this problem has not been solved, or even attempted in practice abroad. The foremost practical electricians of Europe have simply avoided it. To discuss such matters technically is not the purpose of this report; but the Board desires to say emphatically that those fluent critics who talk of putting electric-light conductors underground, *making no distinction between arc lights and incandescent lights, or between the arc lights of different systems,* are ignorant of the alphabet of the subject.

"2. If it should prove, upon thorough and unbiased investigation, that some electric light conductors can be successfully operated underground, while others cannot, the result of an imperative order in this direction would be to drive out of the city the systems which could not go underground. But some systems, of which this might prove to be true, are among the best in all other respects. The characteristically American 'alternating current' for instance, is asserted to give a more uniform light than any other. Before enforcing such distinctions among the different competing systems, the Board must have better opportunity to judge and, to that end, opportunity to examine and test.

"It has been a question whether, and in what circumstances, the electric light companies employed to light the streets under contract with the city could be legally obliged by the Board to place their wires underground. This question has never been a pressing one, because the Board, not having devised a practicable system to include such wires, could not issue such an order, if it had the right. This raises for the consideration of the Board another question, viz.:

"If the city authorities have made contracts at a certain price for lighting the streets, could this Board properly impose upon the contractors an underground system which, though technically practicable, was financially impracticable, under the terms of the contract; that is to say, would inevitably force either the abandonment of the contract or the payment of a higher price by the city. This would certainly bring the Board into an undesirable and mischievous attitude of conflict with the city authorities. Although the letter of the statute * * * leaves the wires of the contractors subject to the authority of the Board, the spirit of the law evidently is, that the city shall not, by the action of the Board, be involved in expense or liability before its constituted authorities shall have fully considered the subject of proposed changes, and appropriated the money to make them. * * * If the Board, by dictating certain conditions to contractors, should force upon the city the alternative of either giving up its experiment in electric lighting of the streets, or paying a much larger sum for that service, the practical effect would be the same as if the Board had ordered the city to give up its fire and police telegraph, or else put them underground at great expense. Furthermore, it is conceivable that such an order as to contractor's wires might leave the city practically no alternative at all but to return to gas. In that case, the question would arise, whether this result would serve the public interest; and it would be a fair inquiry in that connection, whether the citizens of a given locality would rather bear the nuisance of overhead wires than go without the electric light.

"None of the foregoing suggested questions have been completely solved by the Board. The practicability and cost of an underground system of electric light conductors; the exact differences in these respects among different systems employing different intensities and character of currents, and the manner in which, without injury to the public interest, a change to underground conductors can be made, must be determined by further study, and in part by tests and experiments, such as the Board has been unable to make.

"The Board is resolved that it will not stultify itself and bring ridicule upon the law and its beneficent purpose by ordering things to be done, of the practicability of which it is not reasonably assured; that it will seek to carry out the reform intrusted to its charge, in harmony with the city authorities, with the intelligent desires of citizens, and with the business interests of the city, and that it will not be moved from this course by hasty and ill-informed criticisms."

The Commission concludes its report:

"The problem appears to be solved for this city, so far as telephone and telegraph lines are concerned. *With regard to the electric light wires, the Board is not prepared to report at present.*

Prof. Plympton read the following paper before the American Institute of Electrical Engineers, May 16, 1888, on

"UNDERGROUND ELECTRICAL CONDUCTORS IN EUROPE AND AMERICA."

The problems of construction of underground systems of electrical conduction will have been solved when the telephone and the arc light systems are both buried under our streets without impairing the efficiency or durability of either.

By this I mean that all the difficulties encountered in burying conductors are involved in converting telephone and arc light from *aerial* to *underground* systems. Telegraph lines and systems of incandescent lighting present fewer difficulties in the process of burying, and none of a kind not met with in dealing with the systems first mentioned.

The telephone problem is substantially solved. Some details only remain to be settled, among which may be mentioned the best size of conductor, the most serviceable insulation and the maximum distance of effective service for either grounded or metallic circuits.

In Brooklyn the general plan adopted is that of a conduit divided into ducts through which cables containing from sixty to one hundred wires are drawn. The material of the conduit is for the most part creosoted wood. About ten miles of this is already in use in our city, and about four and a half miles of the Dorsett concrete conduit. For the extensions of the underground systems for the present year, only creosoted timber is to be used.

Among the lessons learned from our experience are: 1st. That in creosoted conduits the use of cables covered with kerite, or any similar rubber or gutta percha compound, must be avoided. 2d. That in the so-called lead covered cables, the use of pure lead is also to be avoided, as it is slowly converted into a porous and friable lead carbonate. An alloy of lead with five or six per cent. of tin seems to resist the destructive action. 3d. That the conducting wire first used is too small for satisfactory telephone service. The difficulties of induction and retardation led to complaints as soon as 4,000 or 5,000 feet of underground wire was put in service. The cables that are now being put in are made up of wires whose cross section is greater by one-third than that of the first wire, and they are protected by twice the thickness of insulation.

I mention the above conclusions as having been drawn from our own experience in Brooklyn.

In commencing our work we gathered such information as could be gleaned from localities where solutions of the problem had already been attempted. The impression so largely prevailed that in Europe all telephone and telegraph conductors were underground that our Board decided that a personal inspection of European systems should be made by one of our number. The duty devolved upon me. The result of my inquiry has been published in the scientific papers.

I will briefly refer to a few of the incidents. (Professor Plympton here gave an epitome of the third annual report of Brooklyn Subway Commission, which embodied his observations).

In regard to the burial of the arc light wires I can only say that no method yet tried seems certain of success. Most of them certainly insure the destruction of the underground conductors in from one to three years. But I have no doubt that a solution of the problem will soon be reached, although the system will be kept apart from the telephone and telegraph subway.

It does not seem likely that arc light conductors will be allowed in the same conduit with telephone wires, nor will they be distributed from the same manholes.

In saying that, I believe that a solution of the problem will soon be found. I do not mean to assert that casualties like that recently recorded of a man in the Bowery, who lost his life by grasping the naked wire close to an arc lamp, can be prevented by any system of burying wires. To prevent such *accidents* (if that is the term to be used) the arc lights must be buried with the wires.

All past experience teaches us to proceed cautiously. Nothing can now permanently check the growth of the telephone, the telegraph or the electric light. They have become necessities of our civilization, and any hasty or ill-advised enforcement of the law to convert all aerial to underground systems which should result in serious injury to them, would prove the surest way to perpetuate the nuisances of overhead wires and poles in the streets.

IN BOSTON.

The subject of laying electric wires underground came prominently before the Boston Council in 1886, when the New England Telephone and Telegraph Company, and the Edison Electric Light Company asked leave to lay their wires underground in some of the principal business streets of the city. The Committee on Underground Wires gave two hearings to the parties petitioning, and there were present some of the most expert electricians in Boston, representing the different telegraph, telephone, and electric-light companies. Much valuable information was obtained at these hearings, although, as might be expected, there was considerable divergence of opinion among the electricians. While some of those present were of the opinion that both telephone and incandescent electric-light wires could be successfully operated underground, there were others who held an opposite view, and this was more particularly the case among the representatives of the telegraph companies. It was generally contended that the arc-light wires could not be successfully or safely operated underground, owing to the intensity of their electric current, and the attendant dangers to life and property that would arise in case of imperfect insulation. It was also claimed that the great intensity of the arc-light wires would seriously disturb the working of the telegraph or telephone wires that might be laid in their vicinity.

Notwithstanding this diversity of views, the Board of Aldermen granted the desired permits, and the New England Telephone and Telegraph Company and the Edison Electric Light Company have already laid many miles of underground conduits, in which to place their wires. The former company have now 600 or 700 miles of wire laid underground, in cables containing fifty or one hundred wires, the cables being constructed like the ordinary overhead cables, and each wire being separately insulated. These cables are laid in wooden conduits or boxes, about two feet square, made of creosoted planks, and laid in lengths of about twenty feet; these boxes are subdivided into twelve or sixteen ducts, according to the size of the boxes, and each one of these ducts carries a cable. The junctions of the boxes are insulated and made waterproof by an application of hot tar, with which all the spaces are filled. The conduit is then covered with tarred paper, and an outside casing of creosoted planks. Man-holes are constructed at distances of 200 feet, or nearer if necessary, and from one to the other of these the cables can be drawn through the conduits, a pilot or drawing line being generally left in each duct. Distribution is made by means of lead cables connecting with the conduits. It is the intention of the telephone company to place all of its wires underground as soon as possible in the business portion of the city.

In the fall of 1886, the American Conduit and Construction Company laid a section for the New England Telephone and Telegraph Co., in Tremont street, as an experiment. May 19, 1887, this section was uncovered to see its condition after laying all winter. The following article taken from the *Boston Globe* of May 20, will give the results:

"Superintendent Parker and the engineers of the New England Telephone Company yesterday unearthed and examined the section of artificial stone laid last fall on Tremont street for carrying electric wires underground. The examination was to determine what effect, if any, the wet and freezing weather of the past winter had had on the material. The conduit was found in perfect condition. To determine if the frost had disturbed the joints, a jointed rattan 185 feet long was run through the conduit, and a section of two-inch cable was then pulled through without meeting any obstruction. The engineer of the telephone company expressed himself as perfectly satisfied with the test, and in his opinion the artificial stone has more durability than anything yet tested."

During the past few months a most searching enquiry has been instituted by the Boston aldermen, who were called upon to grant an exclusive franchise to a company to lay down and maintain underground conduits for twenty years. A franchise was granted over the mayor's veto. The usual conflict of statement was met with at every turn.

No arc light wires have been operated underground in Boston. The city has given the company above mentioned the right to lay its conduits in the streets as common carriers, with the power to collect rent and carry all the city wires, but the question is still an open one.

IN DETROIT.

Several years ago, the Michigan Telephone Construction Company laid some Dorsett conduit in Detroit. The Thomson-Houston Electric Light Company were allowed to use a part of the conduit with their wires, but the telephone people soon found their business interfered with by induction and the electric light wires were ordered out. The Thomson-Houston Company also claim that they had numerous burn-outs while their wires were in the conduit.

The common council of Detroit has ordered all wires of every nature within one-half mile circle of the city hall to be placed underground, and this spring the high tension arc light companies propose to try various experiments in the effort to solve the question of practicability.

IN BUFFALO.

The Thomson-Houston Electric Light and Power Company of Buffalo has contracted with the Callender Company for several miles of their solid system of underground wires. The station is not yet completed and no practical test has been made.

IN BROCKTON, MASS.

C. W. PALMER, JR., Superintendent of the Edison Electric Illuminating Company at Brockton, Mass., writes as follows concerning their experience with Callender cables:

"We have here two 'Callender Feeders'; one composed of three cables each 1,262 feet long; two of them being 350,000 c. m. and the third 120,000 c. m. The second feeder is made up of three cables 1,025 feet long; two of them being 120,000 c. m. and the third 50,000 c. m. They cost us $2,300. They were put down last summer and are now giving first-class service, and we expect great things of them in future.

"If you should use this system, be sure and give the placing of it careful oversight, as we experienced considerable trouble shortly after they were first put in use, owing to carelessness in putting the system in the ground. If properly laid, and there is no reason why it cannot be, I regard the Callender system as a first-class thing."

IN DENVER.

In Denver, the city authorities are now laying conduits of a local manufacture, but no arc light wires have as yet been run therein.

Such, in detail, is the test to which arc light wires have been subjected in the United States. In but three cities—Chicago, Philadelphia and Washington—are they actually in operation In Brooklyn, Boston and New York the work of putting them underground has been commenced, and other cities are considering it, but as yet have not used an arc light on the underground cables. Whether an arc light wire can be successfully run underground is susceptible of two conclusions diametrically opposed to each other. It has not been settled beyond question that the general burying of arc light wires is practicable, and, until it is so determined, opportunity for experiment should be allowed by every municipality contemplating ordering the wires underground. It is probable that within a year or two the careful study the subject is having will bring forward some substantial methods, which may then be adopted with confidence and with satisfaction. All other wires should go underground at once.

In this connection, two papers read at the annual convention of the National Electric Light Association held in Pittsburgh, February 21-23, 1888, and a portion of the discussion on them, is of interest.

MR. WELLS W. LEGGETT, President of the Brush Electric Light Company of Detroit, read the following paper on

THE UNDERGROUNDING OF ELECTRIC ARC LIGHT WIRES.

This topic is one of absorbing interest at the present time. Municipal and legislative authority seeks to compel the burying of all the wires, without discrimination, and to cast upon the parties interested, the burden of finding a practical means to accomplish this end. The progress of invention in science and art records infallibly the contemporaneous public demand. This demand discloses a necessity, and inventive ingenuity suggests the remedy.

Starting from this standpoint, we find public wiring began with the telegraph. Marshall of Paisley, in 1753, evolved an electric telegraph, wherein insulated wires were to be trained on poles. Le Lomond, in 1787, Betencourt in the same year; Riezen in 1794; Cavello in 1795; Salva in 1796; Sommering in 1809; Coxe in 1810, and Sharp in 1813, all had telegraphs, employing from one to twenty-six wires, trained on poles. At this stage, however, the interesting experiments, to which the world lent a helping hand, developed signs of commercial utility and value. Man's cupidity and selfishness at once antagonizes what he cannot share, and we find the public arrayed in opposition to the use of the highways. Inventive ingenuity came to the rescue and in 1816, Ronalds erected and used a telegraph, in which some of his wires were placed on poles, and some were buried in the ground. Tribavillet followed in 1828, with a system employing underground wires, and when Professor Morse, in 1832, brought out his system, he proposed connecting Washington with Baltimore, a distance of forty-four miles. To his mind, the wires should be insulated, and laid in a lead tube in the ground. He constructed his line with great care and at large expense; but only a few miles had been laid when the gradual escape from the lines proved his scheme impracticable. He was about to abandon the undertaking, when one of his coadjutors, Dr. Page, of the patent office, or Professor Henry, of the Smithsonian institute, said to him: "Take your wire from the ground and train it on poles." The advice was followed and success achieved. Here, then, in the incipiency of public wiring, is the first recorded failure coupled with its remedy: "Take the wires from the ground and put them on poles."

With underground telegraph and telephone wires the electrical difficulties to overcome are much the same. Moisture must be excluded from the wires. Insulation must be good, and induction must be reduced to the minimum, especially between adjacent lines. In both, however, leakage from the lines, due to imperfect insulation, may be compensated for, within limits, by additional battery force. A battery acts like a pump, sending water through a leaky pipe — the water may all escape before reaching the discharge end — but with a more powerful pump, while increased pressure will cause greater escape through leakage, one may succeed in discharging a limited quantity through the end of the pipe, and so accomplish the purpose sought. For this reason underground telegraph and telephone wires may be operated with a measure of success.

In most of our cities, all telegraph wires might be led to a certain point, from which the wires of all systems could, by one underground conduit, be led to a central station or stations, and thence back through the same circuit, and, where the wires are numerous, this plan might warrant the expense. With telephone companies, whose subscribers are in all parts of the city, the requirements that each wire should be conveyed underground, would involve an outlay for conduits wholly prohibitive.

For direct service incandescent lighting, the required conductors are so large, that to train them upon poles, or to train an equivalent capacity in smaller wires on poles, would involve outlay and annoyance exceeding that required to place the conductors underground. So, again, the resistance of the conductors, to the flow of the current, is exceedingly slight; it might be compared to the pouring of water from a pitcher through a six-inch pipe—its course of least resistance lies right forward through the pipe, and there is little tendency for it to seek an exit through any more restricted channel. In comparison with it the arc light current might be likened to a large hose, with a small nozzle, through which water is being forced with a powerful pump; the tension is very great, and if so much as a pin hole exists in the hose, water will squirt therefrom, and quickly enlarge the orifice to a fatal rent. From the fact, therefore, that telegraph and telephone and incandescent electric light wires may be trained underground with success, by no means does it follow, that the same is true of arc light wires.

It is an undoubted fact, that to successfully underground electric arc light wires, involves simply and solely a question of expense. But expense may be the very essence of the inquiry, for if the expense is out of proportion to the revenue that can be derived from the service, expense alone prohibits the outlay and determines it to be impracticable. Expenditure of money is alone required to tunnel the Dover Straits, and yet French and English capitalists have pronounced it wholly impracticable, as the returns could not warrant the investment. So it is with electric arc light wires; an efficient system, so far as inventive ingenuity has yet presented any plan, involves an expenditure wholly prohibitive, and for that reason alone is impracticable.

All substances are conductors, as copper, rubber, iron, glass, dry air. Some substances are such poor conductors that, in comparison with those that are better, we call them non-conductors. We use these poor or non-conductors, such as rubber, glass, paraffine, wood and dry air, as insulators, but we find dry air to be our most perfect insulator. Dry air is a better insulator than rubber. If, therefore, in dry air, two rubber-covered wires cross in close proximity to each other, the induction is greater between them than would be the case if both wires were bare. But air is usually laden with moisture, and water, charged as it is with mineral salts and acids of the atmosphere, is a good conductor. We therefore coat our line wires with an insulating material, to shield them from direct contact with moist surfaces, or other good conductors, and so reduce in amount the current, which is always escaping by connection to the ground and which varies in direct proportion to the conducting property of the intermediate medium.

When our insulated wires are trained on poles, we have at the rate of 30 poles to the mile, thirty points at which a small amount of current may and always does seep off to the ground. Now, let us put an insulated cable in a conduit in the ground. The conditions are all changed in the most aggravating manner. Instead of comparatively dry air, the air is clammy and heavily laden with moisture; instead of a mass of surrounding air, as on the poles, the envelope of air is, at best but an inch more or less in thickness; instead of resting at 30 points in a mile on glass supports, connected to dry non-conducting wooden poles to make the passage to the ground as much obstructed as possible, the line touches the conduit at say two or three points, in the space of every foot, or say, at ten thousand points in a mile, and the conduit itself, a much better conductor than glass or dry wood, is all that separates it from the moist earth. The bearings at which the current may seep away have been multiplied from thirty contacts to ten thousand contacts per mile, while the conductivity of the medium at each point has been greatly enhanced. A regular arched subway has been suggested, which approaches most nearly to a successful plan, but only a state or city could stand the necessary outlay. A subway of this character, four feet wide and eight feet high, for Chestnut Street alone, in Philadelphia, it was estimated, would cost the city $1,500,000. After completion, such a subway would have to be supplied with forced ventilation, to keep it dry. If, then, the insulated wires could be supported so as to touch at but thirty points to the mile, the conditions of external wire training on poles would be fairly approximated.

The question is frequently asked, if telephone and telegraph wires can be placed underground, why cannot arc wires be disposed of in the same way. The reason is plain, leakage or escape can be supplied, in the former, by an additional battery power, but an arc light generator is a queer specimen of mechanics. It starts a mild current, and this passing back around its field magnets builds them up; the increased magnetism induces greater impulses in the bobbins of the armature; the current, thus increased, continues to pass through the line and back around the field magnets until the current has reached its maximum. It is therefore apparent that anything which saps the current from the line, not only steals the quantity which has escaped, but prevents just so much current from passing back around the field magnets, and to that extent robs the machine of its capacity to generate a current. These machines are capable of taking care of the small amount of loss at the insulators in the pole and line system, but when this is multiplied many times, as in an underground conduit, they are so robbed of their power to recuperate that the resulting lights are necessarily reduced in number and are weak and sickly.

Operations looking to the undergrounding of arc light wires have been prosecuted on a large scale, at New York, Brooklyn, Chicago, Philadelphia and Washington. Notwithstanding all reports to the contrary, I find that at New York, although the subway commission has expended vast sums of money, and has succeeded in burying certain telephone and telegraph wires, no arc light line has, up to this time, been buried in New York City, and this is fully corroborated by the report of the Board of Electrical Control, of January 6, 1888.

At Brooklyn, N. Y., the board of commissioners of electrical subways instituted a thorough investigation of this subject in 1886, and in its report of December 30, 1886, says: "With regards to electric light conductors, this board has found no device which would with certainty in its opinion enable the wires carrying arc light currents to be safely and successfully operated in the same conduit with telephone and telegraph conductors, without disturbance or injury to the latter. * * * But the board desires to say, emphatically, that those fluent critics, who talk of putting electric light conductors underground, but make no distinction between arc and incandescent lights or between the arc lights of different systems, are ignorant of the alphabet of the subject." The president of the board visited all the principal cities in Great Britain and Europe; he found no arc light wires underground, and although a few years since such wires were trained in the Paris sewers they have been removed, and no arc light wires are now allowed therein.

The Brooklyn commission made another report, December 15, 1887, and on this topic says: "As was fully explained in the last report, the subject of underground conduits, for arc currents, is the one which presents the most numerous and novel difficulties. Since it has been impossible for Brooklyn to take the lead in the experimental solution of these, the only remaining course was to watch carefully the progress of experiments in other cities. This has been done, both by correspondence and personal visits of members of the board. The principal cities in which experiments of this kind have been in progress are New York, Chicago, Philadelphia, Baltimore and Washington. In one or another of these, several systems, which were regarded with favor a year ago, have since developed defects or even come to entire failure. It cannot be said that any system has yet been completely proven to be permanently satisfactory. There are one or two, however, which promise well, and this board awaits with attentive interest their further trial; though for the cause specified, they failed to co-operate in the experimental investigation of the different and delicate questions involved." There are no arc light wires underground up to this time in Brooklyn.

At Washington, in 1884, cables were laid in F Street, from Ninth to Fifteenth. In a few months it was necessary to dig up a few of the same. Later there was much more trouble, and it was all taken up and relaid. About one year after the first laying it was wholly abandoned, because they could not make it work. In 1885, they laid cables on Pennsylvania Avenue, Ninth to First Streets, and imported an expert from Antwerp to lay them. Twelve cables were laid in solid cement, and no expense was spared to insure success. They proved an utter failure. The avenue was dug up many times. The wires were in September, 1887, mostly out of service, and the remainder were in such bad shape that they would require constant repairing or have to be abandoned.

An officer of the Washington Company writes me under date of September 12, 1887: "We have many committees coming here to see what we have accomplished, as they have heard that we have met with great success and so forth. We will say to you that our experience, after the outlay of many thousands of dollars, is this: Have nothing to do with underground cables for arc lighting, if you can possibly avoid it; many will tell you that it is perfectly practicable; look out for such persons; they are probably interested directly or indirectly in cables. There is a big lobby in that branch of business. There is no city or town in the world where a cable has been made to work two years, that has been subjected to 2,000 volts of pressure." The Washington Company is using lead covered cables with success for incandescent work; but as to the undergrounding of arc light wires confirmed as late as February 7, 1888, they report that out of fourteen miles of arc light cable, in fourteen different circuits, all have proved total failures and have been abandoned, except a very small amount.

At Philadelphia, elaborate experiments have been prosecuted. All kinds of cables have been employed and a great variety of conduits, but great trouble has been experienced. The systems largely employed were such as had ducts, through which the cables were drawn. In most cases the insulating compound rapidly deteriorated and became useless. In others it would be rotted and become water soaked. When lead covered cables were used, in conduits where creosote was employed, the effect was to rapidly oxidize the lead covering and disintegrate it. It was found that gas could not be kept from the ducts, though apparently gas tight, and many explosions followed. Recourse was then had to ventilation by lamp posts, but the trouble was not remedied. After several explosions in one system, in which persons and property were injured, a power fan was adjusted to force air through the conduit, as it was deemed a better plan than to draw the air through, as by the latter course gas might be drawn in simultaneously. When shortly afterwards the lighting company was congratulating itself that it had overcome the difficulty, a tremendous explosion ensued, extending for a long distance, ripping up the street and breaking a large plate glass window, so that the entire system of lights had to be adandoned. It was some six months ago determined, by the manager, that, as a result of experience, no conduit would suffice for arc light wires in which there were open ducts, and that success lay in the employment of lead covered cables buried solid. This plan, suggested by the city electricians, was tried and success seemed assured, but they now report the experiment unsuccessful. The section buried is not great, but three bad grounds have recently developed. Two were repaired, but the third necessitated the temporary cutting out of a part of the line, and the training of wires on poles, until the earth might thaw and access might be had to the wires. The united companies of Philadelphia are at a loss what to try next.

At this point I would note that these demands for a system for arc light wires have greatly stimulated inventive ingenuity, but yet without success. Out of seventy-two patents that have been issued in this line, fifty-three of them have emanated from Philadelphia, New York, Boston, Brooklyn, Wilmington, Camden and Washington, all in the immediate vicinity of the places where the operations were conducted. The others were mostly from the vicinity of Chicago. Now, as to Chicago,—much has been said of its system,—but a visit to the city satisfied me that its electric arc light industry was being strangled. A few blocks in the immediate heart of the city were using electric lights very lavishly, but the area lighted is scarcely a half mile square. This small area is served from no less than nine separate plants, formerly belonging to as many companies, but now consolidated into one. A letter under date of September 20, from an officer of the consolidated company, explains the situation there; he says: "We have had an enormous amount of trouble with our underground arc light circuits, averaging, I should say, for the last three months, one burn-out every day. The expense of reconstruction and the losses in rebates have been enormous, and the annoyance to our customers is more damaging still. Commencing in April last we bought out some more arc light plants in Chicago, and have proceeded to concentrate them into four main stations, burning about 1,100 lights. Their owners had all been using underground conductors composed for the most part of Okonite, Kerite, Callender and Underwriters' wire. Every-one of these insulators has failed constantly. The only thing that has held up at all is lead covered cable, and we have been driven to the great expense of taking out every foot of the old constructions and of substituting lead covered cables throughout. This work is not done yet. Our people are confident that a large business awaits us here, but its development will depend entirely upon device. These lead circuits may fail much sooner than is anticipated, and it is almost certain that we shall be compelled to reduce the number of lights upon a circuit, a very difficult matter in a city like Chicago. Our conduit space is already filled along the main routes and additional conduits will be necessary. The constant tearing up of the pavements for all kinds of purposes makes our street department very unwilling to grant permits for laying additional conduits, and several times they have refused altogether. I would say here that arc light business in Chicago has developed entirely since the passage of the underground ordinance of 1881. The result has been the multiplication of small isolated plants and the formation of the nine small central station plants bought out by this company. None of these plants made any money. It was impossible for them to do so, operating underground and upon so small a scale. No streets have ever been lighted by electricity here, in fact, the whole industry is in a very backward condition and is likely to remain so, except in the most densely crowded portion of the city, unless some arrangement can be had with the city authorities for overhead wires. The city is trying an experiment of its own in lighting the Chicago river, which will, I am confident, demonstrate the truth of the foregoing statement; and, in the course of a year, I hope that our city fathers will permit overhead lines, in many places where public convenience demands electric lights, where underground construction would forbid it."

At Milwaukee, Wis., three systems have been tried and abandoned, *i. e.*, a wooden duct, plain and tarred, iron pipe and grooved wood. A fourth, consisting of tile conduits with a heavy insulated but not lead-covered cable, has recently been introduced, and is now being tested. The company doubts whether it will stand the trying spring season.

At Detroit, the Thomson-Houston Company employed a cable of the most expensive and approved character, in the Dorsett conduit, and the mechanical work was of the best quality. While the cable was new the results were fair, though loss by leakage rendered it impossible to produce normal light.

It was found impossible to operate telephone wires in the same or adjacent pipes. The company soon abandoned the system, and when its cable was removed, it was found the insulation had so rotted, or softened, that considerable lengths of the wire in many places were stripped bare.

With the alternating system, some wires have been undergrounded in Springfield, Mass., and in Denver, Col., and a small amount in Pittsburgh, but the voltage is only half that of the arc light systems, and even with this low voltage the lead-covered cables have not been in use long enough to determine the question of their success.

The difficulty of explosions from gas has been met with many times at Chicago, New York and elsewhere, as well as at Philadelphia, and seems insurmountable, where gas is employed. At New York men have been suffocated in the man-holes, and in the Western Union building the escape of gas from their conduits has been almost unbearable. At Detroit, an explosion took place in October last, in the middle of the night, in a fire alarm conduit with closed manholes, and in which no wire had ever yet been placed. It doubtless occurred through the admixture of illuminating gas and sewer gas, or other exhalations, in proportions to explode spontaneously. The man-hole cover was thrown high into the air, the street was torn up and the paving blocks were scattered over a distance of eighty feet or more.

The problem of undergrounding arc light wires is by no means solved, but appears to-day to be further than ever from solution, owing to the utter failure of systems, which apparently had all the elements to insure success. In this emergency municipal bodies must suit their action to the facts. To legislate arc light wires beneath the ground, when no practical system is presented for accomplishing that end, is actually and literally to bury the system. When a city is so sanguine of the soundness of its judgment, the remedy would seem to be for the city to provide conduits and the necessary cables, guarantee their success, and then compel lighting companies to rent the lines at a proper figure, or, failing so to do, to quit the field. Summary action on any other basis is incompatible with that justice and equity, which it is the inherent right of every person to demand and receive. Any other course is to discourage enterprise and to ruthlessly impair or destroy capital invested in good faith.

The difficulty usually met with is a peculiar and unreasonable one. Municipal and legislative bodies view with suspicion lighting companies, their officers, stockholders and everybody connected with them. It seems to be assumed that because the companies make a certain showing of facts, the facts must necessarily be exactly the contrary, and they legislate accordingly. Thus, at Detroit, in opposition to letters produced by the lighting company, alleging repeated trials and failures at Washington, Philadelphia and Chicago, the mayor, unquestionably in good faith, reported to the council, as the result of his personal investigation upon the spot, that no electric light wires had ever been buried in the City of Washington, a most glaring error, yet this was followed by a report of the committee and unanimously adopted by the council, that the underground systems of electric arc light wires were entirely practicable and were in successful use at Washington, New York, Philadelphia and Chicago. At Cleveland, recently, the strongest argument in the underground struggle advanced by the city was, that the wires of the Thomson-Houston Company, of Detroit, were operated underground in the Dorsett system, whereas there was not a foot of arc light wire underground in Detroit. To combat this unreasonable and unreasoning prejudice, the question, whether any system yet discovered is practical and sufficient for the purpose, should be thoroughly and exhaustively examined, and an elaborate report made and published, with full data upon which the conclusions are reached; this should be done by a board of competent and eminent men of national repute, entirely disconnected from the electrical business, men who could be gathered from the scientific chairs of our largest colleges and polytechnic institutes, and it would rest with the interested companies of this association to contribute from their own funds, or to influence the necessary contributions by the councils of their respective cities, to pay the necessary expenses of such an inquiry.

At the same meeting JESSE M. SMITH, of Detroit, read the following paper on

"UNDERGROUND CONDUCTORS FOR ELECTRICAL CURRENTS:"

The question of underground conductors of electric currents is one which every person interested in electricity should study fairly and with a predetermination of solving.

It is not purely an electrical question, in fact, the electrical part of it is far overbalanced by the mechanical.

There is no difficulty in finding an excellent insulating material, but the difficulty lies in the holding of this insulator on the conductor.

Many substances are good insulators in dry places; some are good under water; some in damp places; and a few will stand acid and alkaline fumes and the ravages of sewer and illuminating gas, but how many will insulate under all these conditions and in addition be substantial enough to withstand the mechanical injuries to which they are exposed when buried under the streets of our large cities?

A few years ago wire wound around with a little cotton soaked in paraffine was thought sufficiently insulated until the insurance companies had a few losses, due of course to electricity, and then we had underwriters' wire.

The name of this wire has sold hundreds of tons of it, but that does not prevent its grounding a whole system when it comes in contact with the least moisture.

It, however, marks one state in the evolution of the perfect insulation.

There seems to be comparatively little difficulty in making a cable that will carry high potential currents when constantly submerged.

The conditions are substantially the same at all times. There are no alternate changes of moisture and dryness; no great changes of temperature; and very little of the destructive action of gases.

The conditions of the underground conductor are very different.

Here the conductor is dry, then wet; it is frozen, and again thawed; it is attacked by sewer gas, and the corroding action of the water leaching through the accumulated filth of the street; it is subject to the destructive action of the leaking gas and steam pipes, and finally, but not the less surely, to the ruthless "ditch digger."

These are certainly formidable obstacles, but a number of them have already been overcome, and the others must be.

Public opinion says the wires of all kinds must go underground, and electricians, engineers, and capitalists must find the means of doing it.

The question resolves itself into three parts.

1st. The electrical insulation of the conductor.

2d. The protection of the insulator from the effects of moisture and corrosion.

3d. The protection of both from mechanical injury.

The question of electrical insulation seems to me to be solved by at least six of the standard compounds now in daily use.

The second part of the question is the most serious.

If any of the standard insulations can be enclosed so as to be protected from direct contact with moisture their chances of life are certainly improved, but if they can be hermetically sealed, they should be practically indestructible, provided the casing is indestructible.

We single out as among the best materials from which to form such a casing—iron and lead.

Cast iron, underground, will last a great number of years, as shown by gas and water pipes, but it cannot be obtained in lengths much over 12 feet, and the numerous joints multiply the chances of leakage. The conductors must of necessity be drawn into the pipe after it is in place.

The conductor must be considerably smaller than the pipe, and therefore moisture will creep in between the two.

Wrought iron pipe if well coated with asphalt or some similar substance will last a long time.

It may be had in longer lengths than the cast pipe, and the joints are more easily and surely made.

If screw threaded joints are used the conductors must be drawn in, with the same objections as with the cast pipes. If the lengths of pipe are prepared with the conductors in them before laying, joints must be made in the conductor as well as in the casing, at short intervals.

Joints are the bane of electrical construction. From the dynamo to the lamp and return the current is forced to pass joints which frequently offer more resistance than 1,000 feet of wire.

At these joints the insulation, instead of being better than at other points of the conductor, is generally worse, and oftentimes none at all is found.

It is not strange that the current should seek an easier path home, and take to ground rather than be forced through the accumulated resistance of all these joints.

I will venture to say that of all the failures of underground conductors, 90 per cent. are directly traceable to the joints, and for that reason it is desirable to have as few as possible.

The conductor may be obtained in, practically, any length, and the insulation may be put on continuously, but in order to have a continuous casing it must be formed of some soft and ductile metal which can be closed about the insulated conductor in the course of its manufacture.

Lead seems to be the only commercial metal that will meet these conditions, and there are objections even to it.

Lead is soft and easily punctured and offers little resistance to crushing or bending, and is attacked by rats.

On the other hand, the corrosive action of the earth has little effect upon it; while its pliable nature permits of its use in many places where iron pipe could not be used.

Being soft and ductile it can be brought into such close contact with the insulation as to prevent or at least retard the creeping of moisture between the insulation and casing.

It seems to me, therefore, that lead casing offers more advantages and less objection than any other form of protection to insulation as yet open to our use.

The third part of the question, viz: the mechanical protection of the casing, and hence the conductor, is not so difficult a matter.

If an iron casing is used little or no protection is needed.

The pipe is strong enough in itself to withstand any ordinary abuse.

If lead is used, however, it should be kept from contact with sharp stones, bits of glass or metal, and have something about it to warn the "ditch-digger" before he strikes it with a pick.

An ordinary square wooden box is oftentimes sufficient.

If it does rot away it leaves a bed of soft mould.

The box may be made of white oak well creosoted, in which case it will last until the next generation finds something better.

If the underground conductor is to come into practical every-day use, it must be so constructed that it may be tapped at any point, as readily and with as much certainty as a water main is now tapped.

How many miles of lead water pipes are now buried in our streets?

They are not protected by boxes or conduits of any kind, yet it is comparatively rare that we hear of their failure.

If such pipes filled with water can be laid in the earth, without protection, why can not a lead pipe filled solid full with copper and insulation be buried with even more success?

Good workmen can certainly be found who can cut a conductor, splice in a branch and solder it, replace the insulation, and finally wipe a joint on the lead casing with as much ease and certainty as a plumber can make a joint in a lead water pipe and have it stand 100 pounds pressure.

In a general system of distribution there are two classes of conductors, one which we may call through mains or feeders, and another service mains.

A number of through mains may be bunched together in a cable under one casing, where they run in the same direction for considerable distances. These may be drawn into conduits, as they are not to be used between the main distributing points.

Service mains must, however, be so constructed that they may be tapped at any point, and that very readily.

Large sums of money have been spent in New York and elsewhere on conduits for electric conductors.

Are conduits necessary or desirable in electric lighting? Service wires are of no use whatever if placed within them unless man-holes are provided at every 100 feet. Through mains are certainly well protected when drawn into conduits, but at a great expense. No engineer would, I think, at this day, risk putting a conductor in any conduit which he would not be willing to trust under water or laid directly in the earth.

Why not take one half of the money which these conduits cost and use it in buying thicker lead casing and bury the conductor in the ground or in a common wooden box?

The underground conductor of the future, I believe, will be formed with a core of copper cable, thoroughly insulated by any of the standard and well tried compounds; protected by close fitting lead casing, thick enough to resist mechanical injury, and buried in the earth in such a manner that it may be tapped at any point. But where is the money coming from to pay for these conductors?

Whenever electric light plants are installed in a thoroughly mechanical and substantial manner, with due regard to the most economical production of power, so that the electric light is perfectly reliable and obtains public confidence, then the money will be forthcoming.

When capital enough is invested in electric lighting to enable it to cope with the vast sums which have been accumulating these many years in the gas industry, then we shall see our dynamos driven by the waste heat and refuse products of the gas works, and our houses will be heated by gas and lighted by electricity.

The discussion of the two papers just read was as follows:

DISCUSSION.

MR. SUNNY—The statement was made this morning, in the very excellent paper read by Mr. Leggett, that, while it was entirely practicable to work telephone and telegraph wires underground, it was not practicable to do so with electric light wires. I remember distinctly that six years ago, when the question of putting wires underground in Chicago first came up, the telephone interest gathered its experts together, and in meeting assembled they each and every one of them avowed with all sincerity and earnestness, that, while it was possible to work telegraph and electric light wires underground, it would be utterly impossible to operate telephone wires in that manner. The amount of crow that we have been compelled to eat in that time can be estimated, when it is understood that there are two thousand miles of underground wire working in Chicago to-day. As a matter of fact, the whole electrical interest opposed the measure, and left no stone unturned to defeat it. The more they fought, the greater became the general clamor for the removal of the wires. The same thing occurred to which Mr. Leggett referred as having occurred in Detroit,—the statements put forth in good faith and all truthfulness by the electrical fraternity were discredited by everybody on the other side. The authorities exercised their power in its severest form by ordering the removal of every pole and wire from the public streets within a specified time. The prohibition included the entire city of forty-eight to sixty square miles, acres upon acres of which are vacant, miles upon miles of the streets of which are sparsely settled, and alleys in passing through which one instinctively holds his nose, and breathes as little as possible.

The point that I think can be made is, that it is not the best policy to fight a measure in which the public seems to be so greatly interested, but to find a way to give them ten or fifteen per cent. of what they ask for, if, in reason it can be done. I think now, that, had this course been pursued in Chicago, we would now be able to string wires in the air outside of a limit of two or three square miles. Chicago has a thousand arc lights on underground circuits in the center of the city, the greater number of which are in the districts served by the main station located on the Chicago river, at Market and Washington streets. Most of the lighting is east of this point; so that twenty-five circuits, or fifty wires, run practically together for two thousand or three thousand feet.

When this station was established, several months ago, a conduit of twelve creosoted wooden tubes was put down to connect with the Dorsett conduit, built four years ago. The Dorsett conduit is made up of pipes having seven ducts, or holes, two and a quarter inches each. Six lead cables of 3-32 insulation, No. 6 gauge, were drawn into each duct. Lead cables were used in preference to anything else, because of the difficulties experienced in maintaining the insulation where other makes—a good many of them considered first-class—had partly or wholly failed. Many of the conductors having no lead covering have done and are doing good service in various parts of the city; but in the business portion, where the earth is saturated with water-gas and sewer-gas, which circulates more or less freely through the conduit, nothing but lead seems capable of withstanding their influence. The six cables in each duct were made into three circuits,—three positive and negative,—and this arrangement is probably responsible for some of the trouble experienced. It was found that, although the insulation measurement of the lead cables in the conduit was very high,—considerably above a megohm,—and the insulation beyond the point where the lead cables ended, and where some other wire was used, was, say, five hundred thousand ohms, a short-circuit would form, generally within a thousand feet of the station between the two sides of the circuit. The heat generated at the point of the short-circuiting would melt the lead in the adjacent cables, so that in one instance all six wires were rendered useless. Where this occurred in the Dorsett conduit, the destruction did not go beyond the six wires; while in the creosoted wooden tubes, on another occasion, not only were the six wires destroyed, but the inflammable material of which the conduit is constructed resulted in the burning out of the entire conduit and the sixty cables in it. The cold weather of three months had frozen the earth to a depth of four feet; but when the street was opened, it was found that the fire had taken the frost out of the earth below a foot from the surface, and that there was left for a distance of twelve feet a few charred sticks, the bare copper wires, and, in the bottom, some lumps of lead. Iron pipes were put in, and lead cables of 5-32 insulation used, four of which were put in each duct or iron pipe. This heavier insulation, coupled with the redistributing of the circuits, so that positive wires are grouped and kept away from negative wires, will probably result in eliminating this source of trouble.

Looking for the cause of these burn-outs, it was found that they occurred on circuits carrying as few as 20 lights, or 900 volts and 10 ampères, and on circuits carrying 48 lights, or 2,160 volts and 10 ampères.

They generally occurred near the station, although there was an insulation of 6 32 inches between the two sides of the circuit measuring considerably above a megohm; and there were weaker places beyond, indicating that the nearer the station, the greater is the liability of short-circuiting.

It seems incredible that the 3-32 inches insulation around each conductor would not carry this voltage, while the same make of cable of 2-32 carries a greater voltage in another part of the city. We could not, in any instance of a burn out, find that an abnormal resistance had entered the circuit beyond the place where the burn-out occurred. On the contrary, as soon as the damage was repaired, it was found that the circuit was in good order. Without being entirely at rest as to the exact cause of burn-outs, we have adopted what might be called a whip handle method of insulation,—very heavy at the station, and tapering towards the outer end of the circuit. This, it is expected, will enable the conductors to hold the current within them under all conditions.

So much work has had to be done in the main route, on account of the burn-outs, that we have had to make almost constant use of the man-holes in the conduit. We find now that these man-holes are all too small, and must be increased from three and a half feet in diameter to six feet, if possible. The suggestion is made, that the latter size ought to be secured wherever possible, because of the impossibility of properly putting away the labyrinth of wires and cables that must be provided for at these points.

Mr. T. C. SMITH—I wish to state that our company here in Pittsburgh is probably as much interested as any company in the country in aerial wires. We have an enormous amount of pole line already set up in this city, and we are doing immense business from that pole line; and I do not know that we should feel inclined to take the whole of that wire down and go underground within a year, or within two years. But it is very evident that the public demand is urgent that the wires be put underground wherever feasible; and our company, without being asked by the city, and without any ordinances being passed to compel us, has voluntarily undertaken to try the experiment thoroughly. We have seen a great many people try to put their wires underground, and in most of the cases which have come under my personal observation the work has been done in such a manner that it is not a wonder that the cables burned out, or that they ever got current through them. As Mr. Smith, I think, of Detroit, has remarked, it is much more a mechanical question than an electrical one. With regard to one of the conduits of which Mr. Leggett spoke in his paper, the conduit in Chestnut street, Philadelphia, I happened, at the very time that the conduit was being put down there, to be connected with another company in Philadelphia, which was also putting in underground conduits. The underground conduits which our company laid were nothing but a lot of tin tubes put in a wooden box filled with pitch, and we hauled into it first-class lead-covered, insulated cables. We run them for six or seven months without the slightest hitch and without a single burn-out. At the time that the cast-iron conduit was laid in Chestnut street, it was laid without any attempt whatever to keep it water-tight or air-tight. It was nothing but an iron trough in two halves, and they made the absurd attempt to keep it air-tight by putting putty along the edges. When the cables were put in there, I saw a good many being put in, and in some of the cases where lead cables were used, the ends of the cable were simply stripped for two inches of the lead, the conductors spliced, and three or four turns of kerite tape put around it; yet that has been cited as a dismal failure of underground wire. Now, I am not by any means arguing that all wires must go underground, for the simple reason that companies that have spent an amount of money, and have large vested interests in poles, lines and wires, cannot be compelled by any equity to remove the whole of that property and put it underground. In a great many cases it would be equivalent to sending the company out of business. But I do think that companies should endeavor, in the central parts of cities, where the streets are crowded, and the buildings are [high, and there is no doubt that the overhead wires are a nuisance,—that they should make an honest endeavor to put them underground.

With first-class work, I see no reason why underground wires should not be successful. We have heard a great deal about the punching of holes through these cables. Any one who has had anything to do with arc light machinery knows that if you take a forty-light Thomson-Houston machine, or a sixty-light Brush machine, and open the circuit suddenly, you will discharge the field magnets into the line; and if there is any weak spot in the insulation, it will go through it at that instant. We have demonstrated it over and over again. I think, if the gentlemen who are trying underground wires will simply shut their dynamos down regularly, by either short-circuiting the armature or shunting the field circuit that they will do away with a great deal of this punching of holes in the cables. In laying down three or four miles of lead-covered cables, we cannot expect to find them perfect in every part. There will be pinholes occasionally, and the best inspection will not always discover them: but we have no right to assume, because such things as that occur, that the plan is impossible. There is not a man here present who has been in the electric business for a year who has not had a hundred cases of open circuits and short circuits on his overhead lines. The question is not, whether underground wires are feasible or not, but whether we cannot get them established with about the same trouble we took to get our overhead lines in position. I am interested in a company in Philadelphia which is running the underground system entirely. We started with a two hundred volt direct current system. We knew that, if we started that company, the wires would have to be underground; that there were no more overhead ordinances going to be passed. We realized what we were attempting, and we resolved to spare no expense to do the thing properly. When we purchased our cables from the company—they were all lead-covered cables—we told that company to send us their most experienced men for the splicing and laying of those cables. We paid their expenses cheerfully. We not only had them make the joints, but we watched them. We started at the station, and turned the current into those lines. We ran there for six months without a trace of leakage, and then we changed over to the alternating system.

We put a thousand volts into those cables; and they are running to-day, and have been for nine months, without a single ground having occurred on them. We have fifteen miles of cable buried, all told. We have had the question before us, as to whether it would not be possible to run the alternating current in the same box with telephone and telegraph cables. At the request of Chief Walker, the City Electrician, we ran, in the Long-Distance Telegraph Company's conduit, fifteen hundred feet of three-conductor cable. On the two outer wires we put our alternating current, and we placed a twenty-light converter in the chief's office. Into the third wire we connected his telephone, and there is not a trace of interference on it. The telephone company, of course, had given Chief Walker permission to put that cable in their box; but they were a little nervous about the result, being afraid it would knock out their whole system between Philadelphia and New York. After we had been running about a week, they came to us, and wanted to know when we were going to start up. We told them that we had been running about a week, at which they were very greatly surprised; and they have made no objection to the occupation of the box by us. Of course, I have had no real experience with arc lights underground—that is, high-tension currents—beyond the fact that, when I put down my lead cables in Pittsburgh, I use a sixty-five-light Brush machine to test them with. I would not, when testing them, subject them to the strain of opening the circuit suddenly upon them; I do not think it would be policy to do so. Chief Walker, in Philadelphia, is running some four or five miles of lead-covered cable on his arc circuits. He has had one or two burn-outs. He had to string an air.line for one section, because he could not dig up the streets,—being winter. But he did not expect to put down a perfect job the first time.

There is one thing I would like to ask some questions on. It has been stated that a direct-current arc light cannot go underground. I do not think that the gentlemen who make that statement fully realize what a tremendous weapon they are putting in the hands of the press against them. I am not speaking now as an alternating current man, but as a member of the Association, and as representing a lighting company which is interested in overhead wires in so far as they do not interfere seriously with public safety. We have proved beyond any question that the alternating current can be put underground, and can be put in lead-covered cables, without any danger of loss of current. If the direct wires cannot go underground, the alternating can ; and it may be that we shall be called upon to put in new apparatus which we do not want, simply because it can go underground. I think that in a case of this kind it is not the technical press we have to consider; they can realize the difficulty and danger and troubles of going underground; but it is the public press we have to consider. While they certainly do not always express public opinion, they have a good deal to do with molding it; and a persistent attack by the press on a lighting company will compel it to do what they desire. I think the sooner this question of underground wire is faced fairly, the better it will be for all of us. I have had a good deal of consultation at various times with Chief Walker in Philadelphia; and he made the statement to me that some eighteen months ago, before he would recommend the city to spend any money in the experiment, he wished to make a test of a certain cable, and he got a length of three hundred feet of cable. This was cut into two parts, and placed on the two poles,—one on each pole of a Brush sixty-five-light dynamo, laid in the yard of the station with the mud and water all around, and then connected with the air-line. After they had run a few days a fault was developed in the cable, and there was found to be a hole punched right through the insulation. I am not sure, but I think the other connection was found at the end of the wire, where it was bared to attach to the air line. The defective part was cut out, and he has it in his office; and, as I was under the impression that he had been using that wire in that way ever since, I telegraphed him this morning as follows: "Is the three hundred feet of cable originally laid in the yard still in daily operation ?" I received the answer from him, "To the best of my knowledge, it is." And I will ask Mr. Law whether that is the case or not.

Mr. LAW—That cable has not been in use for the last five months at least.

Mr. SMITH—Very good. Then Chief Walker is mistaken. But at the same time he laid a good deal of cable in the upper part of the town, and that cable is still in use. Now, the statement has been made that in Chicago the average capacity of the machines was forty-five lights, and the average circuits twenty lights. I do not know, but I think the man in charge of the station has shown poor judgment in dividing his circuits.

Mr. SUNNY—If you will allow me, I will explain that in one word. We have put out more circuits than the present business requires.

Mr. SMITH—Exactly. I presume that you have done as I have done in Pittsburgh. I have kept all my circuits the size of my smallest unit. But the fact that the circuits and machines do not agree is no proof that it was done because the wire was defective. I would like to call on Mr. Wilber, of the Jenney Company, who has laid a good deal of cable in Philadelphia,—I would like to ask him if his cable is in use; and I would like to ask Mr. Wilber to give his experience.

Mr. WILBER—I would say, for the information of this Association, that last spring we laid over two miles of underground cable in Philadelphia, in the grounds of Girard College. We are running forty-seven lights on two twenty-five-light machines. We have never experienced a particle of trouble with the cable, from the time it was laid up to date. Where the cables came out at the top, and connected with overhead wires, there have been one or two grounds and leaks and lightning troubles, but no serious defect of the cable. From our experience, there is some small grain of truth in the claim made, that you cannot get as great a number of lights out of a given dynamo capacity in running underground as you can on overhead wires. With that exception, I do not know that there is any serious difficulty in running arc-light currents underground, unless it is that of expense, which is sometimes prohibitory.

Mr. LEGGETT—How long is the longest conduit that you have in the Girard College grounds?

Mr. WILBER—It is all one circuit, sir. It is about two miles and an eighth, I should judge. I do not remember the exact length. It is all one circuit, burning forty-seven lights on it.

Mr. LEGGETT—You go out and come back through the same conduit?

Mr. WILBER—Yes, sir, side by side. It is the Standard cable, laid in creosoted boxes, side by side. They come back in the same box.

Mr. LEGGETT—Something less than a mile in length.

Mr. WILBER—The main circuit is about a mile in length. They loop out to the different towers and different places where we have lights.

Mr. LAW—I would like to ask Mr. Wilber if those conduits do not pass through passage-ways that go from one building to another, and if they are not kept at a very high degree of temperature?

Mr. WILBER—I neglected to state, and it is a very important matter, that the buildings of Girard College are heated by steam, and these steam-pipes are carried through a brick conduit. We got the benefit of those brick conduits as far as they went, from the engine-room at the lower end of the grounds up to the farthest end of the building. We ran through this brick conduit, laid our wires in boxes in the conduit, and then branched out to the towers underground,—laying it simply underground. And the duct is always heated in winter up to a very high temperature, and in summer it is not equally warm, because they have a circulation of air by fans, etc., to keep up a circulation of air through the tunnel. But I do not think there is any great amount of dampness in there.

The PRESIDENT—How much of your conduit runs in the duct?

Mr. WILBER—Perhaps from two-thirds to three-fourths of the entire circuit is in this brick duct. Those branches out to the towers run through the ground.

Mr. DE CAMP—Two papers have been read to-day on this subject of underground conductors, one for and the other against; but there is one thing in which they both agree,—perhaps two, but one in particular,—and that is that we must have a perfect insulation. The second is, that it is only a matter of cost. Now, I believe that the first one is true, and no station will be operating a complete underground system until it secures that perfect insulation. In regard to the point that it is a question of cost, I think that ought to be left out until some gentleman is prepared to say just what that cost will be—not that it is only a matter of cost that is keeping the companies from doing it, even keeping them from making a fair, proper, and unbiased test. The question of cost is certainly one of very great importance. Electric lighting from central stations is now nine years old. There are organizations who have back of them probably as good business talent as ever backed up any new enterprise, and, as a rule, they have been organized for business purposes. They are officered and controlled by men who do not hesitate to invest their money in things that they think will ultimately pay,—not to get their money back, because, if that is the case, they have been, without exception, I guess, grievously disappointed. They are waiting for their dividends yet. All know—if they do not they can find out very easily—that the returns upon the capital invested in electric lighting have not been large. They have not been such that would justify the investment of money in a business which is likely to undergo rapid changes or rapid improvements, carrying with them a large depreciation in the value of their property. But the business which they are doing to-day is established; the prices are established upon the basis of the cost of their present plants; and it is unnecessary to say that it would be utterly impossible for any company to go before the public to-day, and advance the price of their lights. The prices of electric lights are high; they always have been, they are high to-day. They are lower than they were when they started, but they are high to-day, and that is the view that every consumer takes of it. Therefore you cannot advance your price. The cost of an underground conductor, those that are said to be entirely reliable to-day, is four or five times as great as that of the aerial wires which we are using, and using successfully. Now, in addition to that, you have got to add the difference in cost in putting those wires down. It will run the cost of your construction up to an enormous amount. That has all got to be provided for. I will anticipate what has been said, and, I presume, will be repeated,—that after you have done that, your expenses cease practically. That I do not believe, from our own experience. I would not venture to say that until I had had a long experience in it. I do not believe it is the case

In regard to the lead-cased cables—I cannot give the reasons for it—there has been a tendency to short-circuit between the wire and the lead. That shows of itself an imperfect insulation. Your lead is no insulator; it is a protector, and it does not protect when your insulation is imperfect. We do not have those troubles with overhead wires. I will state in brief the conditions of the underground wires referred to by Mr. Walker, and I will leave it without any comments as to whether that is a demonstration. I will only go back to the point where I have had the general management of the company that is operating those wires. We start out from the station, and reach the first terminal of the conduit. I will give the figures : 3,375 feet of single aerial wire from station to entrance of conduit; thence by 1,412 feet of single wire one side of the circuit from open conduit; then through conduit 9,500 feet, making a loop of 19,000 feet of lead-cased cable placed in a box six inches square, and filled in solid with pitch. Thence by 30,500 feet of overhead wire one side of the circuit to the station, making a circuit of something less than eight miles, of which about four and a half miles is underground, and part of which is one side of the circuit only. The upper section of that circuit is put in first.

About half of this 9,500 feet was filled in solid lead-cased cable; and I agreed with Mr. Walker, at the time that it was put in, that it was the best I knew of—although I did not know anything about it; it was well made. That cable gave out once. The defect was in the lamp-post, but it was a short circuit between the cable itself and the lead. It was put into the lamp-post carefully, and there was no particular reason why it should be there, so far as anybody could find out. Afterwards a defect in the circuit occurred underground. It being winter time, and the ground being frozen, we strung the wire from one lamp-post to another. There has been since then one defect of the same kind, which has been removed. There is another oné that has been cut out all winter between lamp-posts for the same reason, and strung on poles; it is underground. More recently there is another section between poles cut out in that part of the circuit in the open duct, with one side of the circuit only. In addition to that, we have had what did not concern me, notwithstanding our experience in the Chestnut street conduit; that is a comparatively new section of the city—at least, the gas-mains are comparatively new. Down in the lower part of the city the gas-mains are very old and very defective; but in this part of the city the conduits were comparatively new, and the leakage of the gas was not supposed to be to a degree that would be likely to interfere with the conduit.

There have been three explosions in these man-holes within the last three months. I was favored with being on the spot when one of them occurred. It was in the first man-hole, where the wire came off the poles, and into the side of the man-hole entering the duct. When that man-hole was blown off, it was broken in three pieces. The cover weighed about four hundred or five hundred pounds, I think. In the bottom of the man-hole were three or four inches of water. I happened to notice a little spattering of solder in the water. The man put his hand right under the lead, and he found a small hole where the current had gone through and melted the lead. There was no contact there at all. It came clean out of the man-hole, and went into the duct. Those are troubles that are purely, in my judgment, incidental to the conduit; but there has been another trouble which I think will always exist with lead-cased wire, while it might not with some insulated cable without the lead. That is the difficulty of getting your low terminals safe at the lamp terminal. However, that difficulty would be overcome in some way or other. But it is the question of underground wires themselves. That is the history in this short time of that effort; that is, about eight miles. I just submit whether any company would want to enter into the task of putting down one hundred to one hundred and fifty miles on the strength of that experience; whether they would take the risk of pronouncing that sufficiently successful for entering into a general system. For my part, I would not. It would not take me long to decide that.

In regard to the Chestnut street conduit, which Mr. Smith referred to, that is an iron conduit, and my objection to iron conduits would be pretty much the same as it would be to any lead or any metallic casing outside of the insulation itself. I think the effect would be very likely to be the same. But there is no getting away from the fact, that, as a structure for the convenient handling of wires, the Chestnut street conduit is as good, if not the best, that has ever been brought to my attention. It is certainly a very convenient thing. It is a very expensive thing. As to being well laid, I doubt whether it was; but within the last year that conduit has been nearly reconstructed from one end to the other; it certainly has from Seventh street to Broad. It has been dug up and taken out, and relaid; and the reason it has not been constructed below that is, that we have had a pile of brick on top of it, so that we could not get at it. A blower is constantly kept going to keep that free from gas. The ventilation we have given to it by making connections between the conduit and the iron-lamp-posts open at the top has only been a partial remedy. I telegraphed Mr. Walter F. Smith, superintendent of that station, to be here. I think I am safe in saying that he has had as much experience, if not more, in the putting of wires underground, and the handling of underground wires. In fact, that is his education. He has done very little outside of that. He has been identified with those conduits almost ever since they were started. He is a man of about the ordinary intelligence. He is an electrician. He is a level-headed man; and what he does and has done since he has been managing that station he has done thoroughly and well. Therefore I infer that the work which he has done there has been done as well as anybody can do it. He has made mistakes. He feels that wires can be put underground, but he strikes that snag of expense. He is not prepared to say it could be done at the price at which any company could afford to do it. We had that station, and it was an elephant on our hands. It was not paying; but there was a certain expense which we were subject to, whether that station was running or not. We were running a hundred and twenty-five arc lights from that station, and we needed the income from them.

After the last explosion,—notwithstanding the fact that Mr. Smith thought that the system of ventilating that conduit by a blower was a safe one, we had this last explosion; fortunately the first one that ever did any damage to an extent that we had a claim brought against us,—and after giving the thing proper consideration, I came to the conclusion that the company could not afford to run that station, and take the risk of having to pay such damages, either to property or to life; and we abandoned it, and have converted it into a small incandescent station. Now, in that station we are using a current of a hundred and twenty volts, I think, and it is true with that current we have had no trouble. Our circuits are short, and we have put in the best wire we knew how. In regard to the other explosions which occurred in there, there were seven of them, arising out of the leakage of illuminating gas—I believe it is generally admitted by the committee appointed by the University to investigate it, that it was illuminating gas, and not an admixture of illuminating and sewer gas, and that the explosion was caused by the short circuiting of the wire. Mr. Smith has had some bad wire in there, and he has had some very good wire in there. Now, here is another thing that we ought not to overlook,—that the prices will not be increased, at least, and that there is a much greater prospect of having to decrease them. With the scale of expenses brought as low as we have been able to bring them, with six or seven years' experience to reveal to us our defects and deficiencies in administration, we start out to increase the cost of our construction, without being able to get any compensation for it. The only offset that we have got to that is to save in the mileage of our wire, and it is just exactly what Mr. Smith is doing in making his circuits out to correspond to the smallest machine. Now, we will double up our circuits, and we will double up our machines; or we will make our circuits longer, and double our voltage. That is the only offset that I can see to compensate us for this enormous increase in expense. We have not only got to provide for that twenty-five hundred or three thousand volts which we are using now, but we have got to provide for double or treble that; and I am sure that if this was an experiment which would cost a trifling sum, that would be the thing; but here is an experiment you cannot make except at a very great cost. I have had correspondence, not knowing that Mr. Leggett had had correspondence to the same purpose, and the statements that I get are almost identical with his own.

MR. T. C. SMITH: In speaking of the conduit on Chestnut Street, I was perfectly aware of the circumstances under which the Brush Company took hold of that conduit,—that it was work put down before they got hold of it. Mr. De Camp did not make it quite clear. He seemed to think that I did not consider Mr. Walter F. Smith a competent man. My best answer is that I am using the apparatus which he uses, for putting in the conduits here in Pittsburgh. I think it is because a reliable and conscientious man did the work, that your incandescent lights are running as satisfactorily as they are. If the work that he has done on the low tension cables had been no better than the work done on the old cables, they would have given you the same trouble. I do not think two or three burn-outs on the first cable ever laid has any thing to do with the question of success of underground conduits. It is purely a question of policy for the companies. I do not believe there is a company that has a system of overhead wiring that could put a whole system underground without going into bankruptcy. The main question is, can they not make some endeavor to put the central part of their wiring—in the central part of the city—underground? And by and by I think that they will cut off a great deal of this public opposition to their wires. In Philadelphia, ordinances have been passed allowing overhead wires to be put in the suburbs; but on the main graded streets of the city I think the electric light companies will find their best policy to put some of their wires underground.

Mr. WRIGHT: As representing a company which operates an alternating system, conveying it underground, I would like to make one or two remarks. It seems to me that the matter of underground work has got to be considered. The public will force us to face it. We do not intend to stand just where we are in electric lighting. Electric lighting has come, not only to stay, but to increase; and if we wish to do a larger business, we have got to do a general business; and in a general business we will be compelled to place our wires underground, I think. However, I will simply state to you, gentlemen, that I have been operating for some time an underground system. It is not very extensive, and it has not been operating for a great length of time,—some four months; but during that period, and possibly during the worst part of the year, —during the fall of the year, when we are troubled with water in our conduits and man-holes,—we have had no trouble whatever in that conduit, and the only occasion on which I have had to open man-holes for the purpose of getting to my conduit were due to outside causes. Our buildings are connected directly at the man-holes and junction-boxes; and on two occasions, when my men were connecting converters, they short-circuited and blew out the safety fuse, and on these occasions only we opened the conduit. When I left Springfield we had fifteen inches of packed snow on the streets. Our company do not pretend that they were doing a big thing in doing the underground work. It was simply because they were compelled to. We have no poles there. All our arc wires are on structures on the buildings. It was impossible for us to do any work there unless we went underground. I think it will be found in the next year or two that a very large number of companies in this country will be placed in exactly the same position; and, that being so, we must all look to the future, and try and learn all that we can about underground work. My advice to our company would have been to wait and let other people experiment, had we not been compelled to go underground. However, we are running underground. The work is not costing a great deal of money, and I do not expect that we will have a great deal of trouble. I simply state that we are doing this as a proof that it can be done, for a short time, at all events; and if it can be done for a short time, it can be done for a long time.

SYSTEMS OF UNDERGROUND LINES.

Inasmuch as this question of underground wires must soon be met in all the large cities, it is well to consider what system should be adopted. The principal systems of electrical subways may be divided, first, as to their material composition, and second, as to their mechanical construction, and the manner in which the wires are placed in them.

As to material composition, subways are: first, of insulating material, such as wood, glass, concrete, etc., etc.; second, of conducting material, such as iron.

As to mechanical construction, subways are, generally speaking: first, tunnel systems; second, "drawing-in" systems; third, solid systems; fourth, "dropping-in" systems; fifth, combined systems.

TUNNEL SYSTEMS, or those where space is provided underground sufficient to allow the passage to and fro of men who place the wires within the subway, could be recommended, were unlimited time and money at the disposal of the city or the private company; but the expense of such a system precludes the adoption of such a plan in general. If ever underground railroads become a feature of city transportation, then, perhaps, the tunnel can be used for some of the future trunk line cables. In Paris, where the foundations of the city are honeycombed in all directions by large sewers, such a plan is practicable and admirable, but it is not to be thought of in most of the American cities.

DRAWING-IN SYSTEMS, or those where man-holes are provided in the streets, connected by tubes or pipes through which the wires can be drawn, are next in prominence and convenience to tunnel systems.

SOLID SYSTEMS, or those where wires are permanently embedded in insulating material and incapable of being reached except by tearing up the streets and the insulation, have been found to work with more or less successful. Many electricians prefer this kind of underground service.

In the future, when the uses of electricity become more general, and its nature better known, it may be that some cities will require a grand electrical underground highway, where space can be provided for the conducting and distributing of sufficient power to run all the engines and work all the machinery within the city limits. It may be and probably will be, that before such space is needed much more will be known of the qualities of different forms of electrical conductors, and of the best methods of carrying them underground, and that the matter of electrical subways will be no longer experimental, but practically demonstrated in every detail.

Leaving out of consideration all tunnel systems as too expensive, cities should also discard any system which calls for the simple laying of insulated cables in the earth. They would not stand the chemical action of the gases and acids; the streets would be continually torn up for new connections and repairs. Municipalities are thus shut up to the question of electrical subways or conduits, or solid material in which the wires or cables, insulated or otherwise, must be placed, and which once laid down should meet all the demands of the present and near future.

MATERIAL.

As regards a conduit, the principal requirements are that it be cheap, water-tight, have sufficient capacity, permit (without disturbing the pavement) the insertion, removal and addition of wires, and allow access for repairs and testing. The material for, and construction of, a conduit are, in a measure determined by the particular system of conductors it is to carry.

Various material have been suggested for an underground conduit—iron, wood creosoted or coated with paraffine or asphalt, glass, porcelain, earthenware coated with asphalt, stoneware, artificial stone, glazed terra-cotta, asphalt-coated terracotta, paper saturated with asphalt, and various compositions, such as marble dust and asphalt; sand, linseed oil and asphalt; sand, broken stone and asphalt; sulphur, sesquioxide of iron and bitumen; sand, paraffine, asphalt, with the addition of certain chemicals, rosin and other ingredients, etc., etc.

As far as durability is concerned, wood properly treated, is known to last underground a great many years, and could be used in the construction of a conduit. Creosoted wood is, however, injurious to gutta percha covered wires. Earthenware or terra-cotta covered pipes have been used in England, and if glazed or coated with asphalt and well jointed to exclude moisture, could be used for underground service. Glass and porcelain are expensive materials. Asphalt saturated paper could be used for the lining of passages in some forms of underground conduit. Artificial stone moulded into a continuous conduit is cheap, and one of this kind was laid in 1881, in Market street, Philadelphia. Some is to be laid in Boston. In this conduit tin tubes covered with asphalt are imbedded in the artificial stone.

Terra-cotta, thoroughly coated with asphalt, is a cheap material, and when protected by an outer box or covering of creosoted wood is adapted for an underground conduit of the usual form. It is unaffected by changes of temperature, there is little or no difficulty in securing and maintaining tight joints, and if properly coated with asphaltum, it is indestructible.

An idea of the extraordinary complication in which this whole matter of selection is involved, may be gained from the statement that there are nearly 600 different plans of electrical underground conductors.

Solid conduits are claimed to be satisfactory in some cities and are being generally experimented with. When once a wire or cable so bedded becomes imperfect, it must be abandoned and can never be replaced without digging up the whole bed, and no new wires or cables can be introduced when needed without laying a new bed or reopening the streets, so that the owner must insert, once and for all, the whole number of conductors he thinks he may need at any future time, thus incurring a great and oftentimes unnecessary outlay of capital.

The laborer's pick has been one of the great objections to any cement or tiling system, and it is now nowhere seriously considered, except it be placed under the sidewalks.

The New York Board of Electrical Control, after a careful examination of the plans submitted, and after considering all the facts, recommended an asphaltic concrete; but it also recommended *the most rigid testing and inspection* of the conduit while being manufactured and laid, and the exacting of guarantees from the manufacturers that their work would stand certain tests. In a certain sense municipal bodies can afford to be liberal in allowing the electric companies to use any kind of insulated conductors they prefer, but certain rules should be laid down that will insure the exclusion of untried and experimental compounds, which, should they fail, will cause the frequent reopening of conduits and cause inconvenience and trouble.

WHAT WILL CITY COUNCILS DO?

If a city should decide that arc light wires can be successfully operated underground, the questions to be met are, will the city compel the several electrical companies to put their wires underground indiscriminately; will it exercise supervision over the private conduits; or, will it build a conduit into which its own wires and the wires of all companies can enter on payment of a rental.

In a general way there are four parties in interest in this matter—the people, the city, the companies operating electrical conductors, and the company which builds the conduit and rents it, if such a plan should be adopted by concurrence of the electrical companies. The interest of the people is identical with the duty and interest of the city, in that the wires shall be placed underground speedily, safely, and without deteriorating from the efficiency of the electrical service as it is at present. The interest of the companies is bounded by the cost and trouble of changing from an overhead to an underground system, of maintenance when so changed, in securing such facilities as they may need on fair and impartial terms, and in being guaranteed against interference with their conductors and the impairment of their efficiency. The interests of the construction company, if one there be, is in securing a fair return for the investment made in carrying out the plan of the city.

The cost of establishing the present overhead-wire system has been excessive, and this expense will be practically a loss to the companies when the system is abandoned and the wires are placed underground, so that, in considering the subject equitably, city authorities should have an eye to the protection of the capital invested under their jurisdiction, as well as to the safety of life and property.

Addenda.

SINCE this book has been in press the suit at law referred to on page 112 has been determined by the decision of Judge Gresham, who held adversely to the claims of Mr. Brush and in favor of the Jenney Company, which was the defending party. The case will doubtless go to the Supreme Court of the United States.

Another important judicial decision affecting electric lighting interests has been given since this book was placed in the printer's hands. It is that of Judge Wallace, in the United States Court for the Southern District of New York, and affects the carbon filament in the incandescent lamp. The court held that Mr. Edison had allowed his foreign patent to lapse and that therefore he could not prevent the United States and other companies from making the lamps. This case will also doubtless go to the Supreme Court for final adjudication.

Frederick, Md., has decided to purchase the Thomson-Houston plant located there at a cost of $14,500.

The Board of Gas Trustees of Wheeling, W. Va., have recommended that a plant be purchased at a cost of about $23,000.

Urbana, O., has contracted with the Fort Wayne Jenney Company for an arc and incandescent plant.

Edison street lamps are used in Oskaloosa, Ia., Peru, Ill., Olney, Ill., Monmouth, Ill., Seattle, W. T., Tacoma, W. T., Attica, Ind , Pendleton, Ore., and Brockton, Mass., in addition to the cities named in the contract system.

Beaver Dam, Wis., Conshohocken, Pa., Peekshill, N. Y., Plainfield, N. J., Salem, O., Stapleton, N. Y., Torrington, Conn., Tyler, Texas, and Tyrone, Pa., are lighted by the Westinghouse system of incandescent lamps.

Index to Advertisers.

	PAGES
American Conduit and Construction Co., Boston	19
American Tool and Machine Co., Boston	24
Ansonia Brass and Copper Co., New York	16
Babcock & Wilcox Co., New York	237, 238
Bishop Gutta Percha Co., New York	10
Brady, T. H., New Britain, Conn	253
Brush Electric Co., Cleveland	255
Callender Insulating and Waterproofing Co., New York	12
Campbell Electrical Supply Co., Boston	244
Central Electric Co., Chicago	2
Clark Electric Co., New York	5
Cleveland, W. B., Cleveland	242
Cleverly Electrical Works, Philadelphia	21
Detroit Electric Tower Co., Detroit	254
Eastern Electric Cable Co., Boston	16
Eclipse Wind Engine Co., Beloit, Wis.	16
Electrical Accumulator Co., New York	248
Electrical Construction Co., Chicago	15
Electrical Review, New York	241
Electrical Supply Co., Chicago	cover
Excelsior Electric Co., Chicago	236
Fort Wayne Jenney Electric Light Co., Fort Wayne, Ind	17
Greeley, E. S. & Co., New York	23
Globe Gas Light Co., Boston	234
Harrisburgh Car Mfg. Co., Harrisburgh, Pa	250
Hartford Dynamic Co., Hartford, Conn	12
Heisler Electric Light Co., St. Louis	229-232
Heine Safety Boiler Co., St. Louis	6
Holmes, Booth & Haydens, New York	15
Hooven, Owens & Rentschler Co., Hamilton, Ohio	8
Hussey Re-Heater Co., New York	22
Jarvis Engineering Co., Boston	20
Jenney Electric Co., Indianapolis	cover
Lane & Bodley Co., Cincinnati	245
Light, Heat and Power, Philadelphia	246
Main Belting Co., Philadelphia	25
Mather Electric Co., Manchester, Conn.	14
Modern Light and Heat, Boston	234
Moore, Alfred F., Philadelphia	19
Muckle, M. R. Jr. & Co., Philadelphia	248
Munson & Co., Chicago	239
National Iron Works, New Brunswick, N. J	233
National Pipe Bending Co., New Haven, Conn.	21
New York Electrical Construction Co., New York	256
Non-Magnetic Watch Co., New York	240
Noye, John T. & Co., Buffalo, New York	252
Otis Bros. & Co., New York	10
Parker-Russell Co., St. Louis	16
Payne, B. W. & Sons, Elmira, New York	3
Pearson, W. B., Chicago	13
Pond Engineering Co., St. Louis	11
Queen, Jas. W. & Co., Philadelphia	234
Rider Garbage Furnace Co., Pittsburgh	251
Russell & Co., Massillon, Ohio	5
Santley, W. R. & Co., Wellington, Ohio	5
Sawyer-Man Electric Co., New York	18
Schieren, Charles A. & Co., New York	23
Schuyler Electric Co., Middletown, Conn.	24
Smith, Jesse M., Detroit	244
Solar Carbon & Mfg. Co., Pittsburgh	248
Standard Carbon Co., Cleveland	242
Standard Underground Cable Co., Pittsburgh	3
Stellar Electric Co., Boston	244
Tatham & Bros., Philadelphia	243
Terre Haute Electric Light & Power Co., Terre Haute, Ind	235
Thomson-Houston Electric Co., Boston	7
United States Electric Lighting Co., Chicago	1
Van Depoele Electric Mfg. Co., Chicago	4
Waterhouse Electric & Mfg. Co., Hartford, Conn	9
Western Electric Co., Chicago	cover
Western Electrician, Chicago	247
Westinghouse Electric Co., Pittsburgh	249

MUNICIPAL LIGHTING

By Low Resistance Incandescent, of any Desired
Candle Power, Connected in direct Series.

HEISLER SYSTEM.

It is an acknowledged fact among Experienced Electricians and is beginning to be generally recognized by Municipal Authorities and the people at large that Street Lighting can be done much more effectively by the Incandescent System than by Arc Lighting, for the reason that while the Arc Lamps light up a limited area more brilliantly than the Incandescent Lamps do, the latter, by reason of the great number that can be maintained at the cost of an Arc Lamp, will afford a much better distribution of Light.

The comparative cost to municipalities of Arc and Incandescent Lamps being 8 or 10 to one, it needs no argument to demonstrate that from eighty (80) to one hundred (100) Incandescent lamps would give a very much more uniform distribution of light over a given territory at the same cost than ten (10) Arc lamps.

If one hundred (100) Incandescent lamps were placed two hundred (200) feet apart in a given territory, the Arc Lamps that could be obtained at the same cost to cover the same area would have to be placed from sixteen hundred (1600) to two thousand (2000) feet apart.

The glow lamps of the Heisler System, by reason of their superior brilliancy are particularly adapted to Street Lighting, as they are, so far as brilliancy is concerned, small Arc Lamps.

In cities having their streets lighted by gas, the difference between the amount of illumination obtained at the same cost from gas lamps and the Heisler Glow Lamps is still more marked.

One Heisler Lamp of thirty (30) C. P. will light a street intersection much more brilliantly than two ordinary gas lamps, while the one Heisler lamp can usually be furnished at or near the cost of a single gas lamp, as they are usually sold in small cities.

The practically unlimited distance to which the current can be carried by the Heisler System renders it available for lighting country towns and sparsely settled districts, where Electric Lighting by any other system would be simply impossible.

The following estimate of the cost of a plant with a capacity of five hundred 20 C. P. lamps has been prepared from a plant actually put up, and in connection with the statement of the income that it will earn (based on a scale of charges below the average) and the accompanying estimate of running expenses, shows what may be done in towns that can only afford a plant of moderate cost, and will prove interesting reading to investors.

Estimated cost of installing a plant with a capacity of 500-20 C. P. lights with seven miles of Street and Commercial circuits and steam plant complete:

Electrical apparatus set up complete	$6,500 00
Line wire put up complete and all connections made	1,650 00
Steam plant complete	2,500 00
Total	$10,650 00

Add to above, cost of ground and building for Central Station, which will vary with each locality. Cheap ground, however, can be used as the station can be established in the outskirts of a town where land is cheap, or *water power can be utilized if miles away*.

The above plant will earn, at rates below the average, as follows:

160 Street Lights at $20 per annum	$2,000 00
300 Commercial Lights to 10 o'clock, at $15 per annum	4,500 00
100 Commercial Lights to 1 o'clock, at $18 per annum	1,800 00
Total	$8,300 00

A liberal estimate of the cost of running the above plant per annum will be as follows:

Wages—Engineer, Lineman and Superintendent	$2,000 00
Fuel, oil and waste	850 00
Renewals of lamps	700 00
Taxes, insurance and incidentals	200 00
Total	$3,750 00

HEISLER ELECTRIC LIGHT CO.,

809-817 South Seventh Street,   -   *ST. LOUIS, MO.*

The above map shows the distribution of Incandescent Lights on the streets of the Villages of Matteawan and Fishkill Landing, on the Hudson River, 50 miles from New York.
Eighteen miles of streets illuminated.
One hundred and sixty 16 candle power Gas Lamps, costing $30 to $34 per post per year, have been replaced with Incandescent Lamps of 20 and 30 C. P. at $20 to $25 per year apiece.
This plant was started on the first of September, 1887, with 160 street lamps, which proved so satisfactory to the Municipal Authorities that they soon gave an additional order for 50 lights to the owner of the plant, Mr. William Carroll.
On the 22d of December, 1887, Mr. Carroll started a commercial circuit and soon had all the lights contracted for that a second and third machine would carry.
The motive power for 450 30-C. P. lights is supplied from one Turbine wheel of 65-H. P.

HEISLER ELECTRIC LIGHT CO.,
800-817 South Seventh Street, ST. LOUIS, MO.

Municipal Lighting

By Low Resistance Incandescent,

Of any Desired Candle Power,

Connected in Direct Series.

HEISLER SYSTEM.

Reasons why the Heisler System of Electric Lighting
is Superior to all Others.

Combining successfully the illumination of streets with the universal supply of incandescent light to every private house, suitable to comply with all the various demands of commercial and domestic life by the most perfect Automatic regulation, with every facility for changing and shifting the circuits, or extending the same to any desired distance, at very small cost, this system has made Central Station Lighting a practical success.

The remarkable financial success that has attended the adoption of the Heisler System proves this; twenty-four Central Stations erected in one year, all being on a paying basis.

Every single plant erected so far on the Heisler System is a success financially, and making money. This can be truthfully said of no other system.

The electrical current can be carried to any desired distance, with the loss of only one 30-C. P. light to every three ½ ohms of line wire resistance (about one 30-C. P. light per mile) This makes the length of line practically unlimited. Circuits of fifteen to twenty miles, and even more, having been constructed for lighting the streets.

Water power located miles away can be used to advantage, the power in some of our plants being located over five miles from the lamp distribution. This is simply impossible with any other system.

It is the only practical mode of Central Station Incandescent Lighting, combining the illumination of streets, stores and dwellings with 15, 20, 30, 45 and 60 C. P. lamps on a single wire.

The machines require very little attention, do not spark nor heat, and run at a speed of only 600 to 800 revolutions. No extra high speed engines are therefore required.

Our system is not multiple series nor multiple arc, and lamps are not run in groups, as is the case in multiple series incandescent lighting circuits. Any single light can therefore be turned on or off without turning on or off the whole group.

The lamps are connected on one wire in series. Shifting, changing, or extending the circuits is therefore simply a matter of splicing on additional wire. No expert is needed to make mathematical calculations on the size of the conductors, and any ordinary workman can be taught to handle our lines successfully.

Any given number of lamps can be placed with economy at any point along the circuit.

The farthest lamp from the dynamo burns just as bright as the lamps in the station.

The candle power of our lamps remains constant and uniform. The lamps give a brilliant white light, remain perfectly bright during their whole life-time, and the globes do not get blackened on the inside as all the high resistance lamps do.

HEISLER ELECTRIC LIGHT CO.,
809-817 South Seventh Street, - ST. LOUIS, MO.

Municipal Lighting

By Low Resistance Incandescent,
Of any Desired Candle Power,
Connected in Direct Series.

HEISLER SYSTEM.

Reasons why the Heisler System of Electric Lighting is Superior to all Others.

We guarantee the life of the lamps to average 800 hours. As a matter of fact their life runs from 1,000 to 1,200 hours, as shown by reports from parties using our system. The report of the St. Louis Illuminating Co., for March, 1888, gives the average life of lamps at 1,135 hours.

Our light commands a higher market price than any other, being perfectly white, of great brilliancy and unvarying steadiness, and is much the superior of any other Incandescent light.

Our system is not hampered with those numerous devices which complicate the service without ever fulfilling the purpose for which they are made, such as safety-catches, shunt boxes, converters, accumulators, distributors, etc., as none of them are necessary to it.

Seven to thirty C. P. lights, or two hundred and ten (210) C. P. to the H. P. is guaranteed, being about one-third more than is obtained by any other system of Incandescent lighting. This insures much greater economy of operation.

A single wire only of No. 8 B. and S. gauge is used for main line circuits. This insures greater economy of construction.

It is a low resistance system, the lamps requiring an Electro-motive force of only 10 to 12 volts, as against 50 to 100 volts of the high resistance system. This insures economy of power.

The dynamos can be handled with impunity when running under a full load, there being only 45 volts at the brushes. This insures safety in handling.

The Automatic Regulator is connected directly with the steam plant. It is consequently powerful and reliable under all conditions of loading. It is entirely automatic, the Regulator adjusting the current exactly to the number of lights burning in the circuit, without any attention whatever. This insures against carelessness of attendants.

In judging of the candle-power per H. P. it should be borne in mind that other systems have to insert resistance whenever lamps are turned out or filaments broken. We reduce the current itself in accurate proportion to the number of lights on the machine.

The power required is therefore directly proportioned to the number of lights in use, and is reduced as lights are turned out. This saves coal.

The principal plants of the Heisler System are at the following places:

St. Louis, Mo.	Leavenworth, Kans.	Ottawa, Kans.
Ogden City, Utah.	Pendleton, Ore.	Matteawan, N. Y.
Eugene City, Ore.	Napa, Cal.	Saugerties, N. Y.
Albany, Ore.	Wabash, Ind.	Orange, N. J.
El Paso, Tex.	Johnstown, N. Y.	Vincennes, Ind.
Monticello, Minn.	Salt Lake City, Utah.	Mankato, Minn.
Liberty, Mo.	Kingman, Kans.	Fergus Falls, Minn.
East Portland, Ore.	Salem, Ore.	Red Bank, N. J.
Fayetteville, Ark.	Ocean Grove, N. J.	Hackettstown, N. J.

HEISLER ELECTRIC LIGHT CO.,

809-817 South 7th Street, - ST. LOUIS, MO.

WATER TUBE STEAM BOILERS

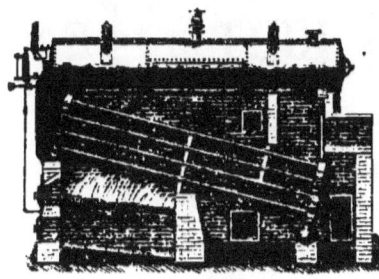

Safe
From Disastrous Explosions

Economical
In Every Respect.

Durable
In All Their Parts.

EASILY CLEANED and EXAMINED.

(MOORE'S SYSTEM.)

OUR Special Method of Construction enables us to erect our Boilers in places, when, on account of limited space, TUBULARS, or those of other styles, COULD NOT BE ERECTED.

References and further Information furnished on Application.

MANUFACTURED BY THE

National Water Tube Boiler Co.
NEW BRUNSWICK, N. J.

The STANDARD ROCKING GRATE BAR
(PATENTED.)

SOLD On their Merits.

ENDORSED By every Customer.

| EVERY LEAF REMOVABLE FROM THE MAIN BAR. | Air Space Adjusted for ALL KINDS OF FUEL.
 WRITE FOR CIRCULARS, REFERENCES AND PRICES TO
 NATIONAL IRON WORKS
 New Brunswick, N. J. | GREATEST ECONOMY IN THE USE OF FUEL. |

The only Weekly Electrical Newspaper in New England.

MODERN LIGHT & HEAT

PUBLISHED EVERY THURSDAY.

Devoted to the interest of Electricity, Gas, Fuel Gas, Heat and Power.

OFFICIAL ORGAN NEW ENGLAND ELECTRIC EXCHANGE.

Subscription, - - - - - $3.00 per Annum
" Foreign Countries, 4.00 " "

Advertising rates furnished on application, address

MODERN LIGHT & HEAT,
178 Devonshire St., **BOSTON, MASS.**

STANDARD ELECTRICAL TEST INSTRUMENTS.

Ayston & Perry's New Spring Ammeters and Voltmeters, Edelmann, Hartmann & Brown's Galvanometers, Bridges, Rheostats. Also, by all the prominent makers. Call and examine.

JAS. W. QUEEN & CO.
924 CHESTNUT ST., PHILADELPHIA.

Isaac Stebbins, Prest.   S. Wheeler, Treas.   D. W. Lee, Gen'l Mgr

Street Lighting by Contract

The Globe Gas Light Company
OF BOSTON.
Incorporated 1874.

Contractors for Lighting Streets of Cities and Towns

Constantly on hand their celebrated GLOBE NAPHTHA in barrels and cans. Also full assortment of MILL, LAWN and STREET LANTERNS, LAMP POSTS, FLAMBEAUX, or VAPOR TORCHES to VAPORIZE OIL, NAPHTHA or GASOLINE.
LANTERNS FOR HOTELS AND PRIVATE GROUNDS.
Please call and Examine.

Office and Warerooms, 77 & 79 Union Street, BOSTON.

Terre Haute Electric Light AND POWER CO.

MANUFACTURERS OF
HAMMERSTEIN'S
—ADJUSTABLE—
Lamp SUPPORT AND ARM
(Patented June 29, 1886.)

It can be adjusted to vary the height of the lamp from the ground. The method of operating is so simple and speedy that one man can trim 100 double arc lamps in a day. We have 210 of these in use in lighting the streets of this city, covering a territory of 35 squares north and south, by 20 squares east and west. Two men cover this ground and trim the entire 210 double lamps every day. They extend the lamp 22 to 28 feet from the street corner, thus placing the lamp at the intersection of the four corners, giving good light in four directions. Shady streets are no obstacle where these supports are used, as the lamp can be set low, if necessary, and throw the *light under the trees*. They can be used under any line of telephone or telegraph wires with 20 inches of space above. They *will not freeze up in sleety or snowy* weather as is the case when blocks and pulleys are used for suspending street lamps. No step ladders or windlass used. We believe them to be the simplest, the safest and the best on the market. Prices quoted on application. Address

TERRE HAUTE ELECTRIC LIGHT & POWER CO.,
TERRE HAUTE, INDIANA.

Reference.—City Terre Haute, Terre Haute, Ind.; Brazil Electric Light Co., Brazil, Ind.; Crawfordsville Gas and Light Co., Crawfordsville, Ind.; Washington Gas Co., Washington, Ind.; Brush Electric Light Co., Owensboro, Ky.; Twin City Gas Co., La Salle, Ill.; W. F. McKinney, Champaign, Ill.; Merchants' Electric Co., Danville, Ill.; Belvidere El. Co., Belvidere, Ill.

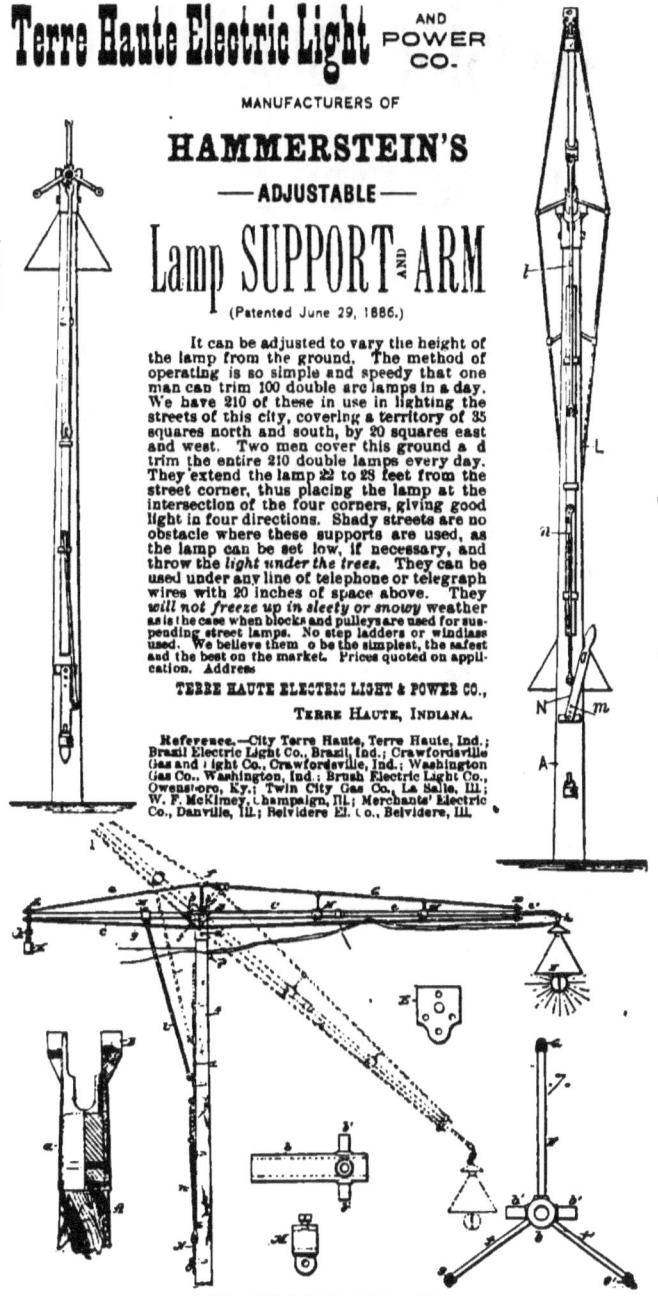

AN ARC LAMP SUPPORT.

Excelsior Electric Co.

HOCHHAUSEN SYSTEM.

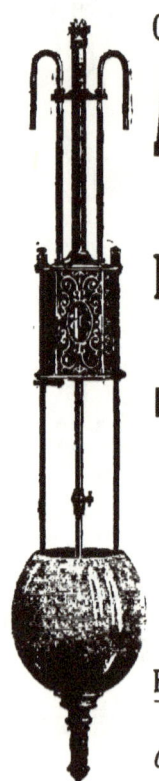

COMPLETE SYSTEM
—BOTH—
Arc and Incandescent
ELECTRIC LIGHTING.

Municipal Lighting
A SPECIALTY.

PERFECT
AUTOMATIC
REGULATION

Guaranteed to cut down to one Light, saving power in proportion to the number of lights burning.

Lights Perfectly Steady.
Free from Hissing and Flickering.

Catalogue and Full Information
FURNISHED ON APPLICATION.

EXCELSIOR ELECTRIC CO.
11 East Adams Street,
CHICAGO, - - ILL.

THE BABCOCK & WILCOX BOILER
In Electric Lighting Stations and Isolated Plants.

THOS. A. EDISON, Orange, N. J.,				Aug., 1887,	3	219
EDISON MACHINE WORKS, Schenectady, N. Y			1st order,	Jan., 1881,	5	448
do	do		3d do	Sept., 1887,		
EDISON ELECTRIC LIGHT CO., Paris, France,				June, 1881,	1	150
do	do	London, Eng.,	1st order,	June, 1881,	2	300
do	do	do	2d do	June, 1882,		
do	do	Milan, Italy,	1st do	Aug., 1882,	8	1312
do	do	do	5th do	Feb., 1887,		
do	do	Livorno, Italy,		Sept., 1887,	3	438
EDISON LAMP CO., Newark, N. J.,			1st do	June, 1846,	3	281
do	do		2d do	July, 1886,		
EDISON ELECTRIC ILLUMINATING CO., New York,			1st do	Sept., 1891,	14	2620
do	do	do	3d do	Dec., 1887,		
do	do	Lawrence, Mass.,	1st do	Sept., 1882,	3	286
do	do	do	3d do	April, 1884,		
do	do	Fall River, Mass.,		Oct., 1883,	2	146
do	do	Shamokin, Pa.,		June, 1883,	2	146
do	do	Hazleton, Pa.,		Nov., 1883,	1	92
do	do	Bellefonte, Pa.,	1st order,	Nov., 1883,	2	184
do	do	do	2d do	May, 1885,		
do	do	Tiffin, Ohio,		Nov., 1883,	1	92
do	do	Middletown, Ohio,	1st do	Dec., 1883,	2	143
do	do	do	2d do	Aug., 1884,		
do	do	Piqua, Ohio,		March, 1884,	1	92
do	do	Circleville, Ohio,		April, 1884,	1	92
EDISON CO., FOR ISOLATED LIGHTING, Washington, D. C.,				April, 1884,	2	164
do	do	N. Y. Herald, N. Y. City,		Nov., 1861,	1	150
WESTERN EDISON ELECTRIC LIGHT CO., Chicago, Ill.,				July, 1882,	1	40
U. S. ELECTRIC LIGHT CO., New York,			1st order,	Feb., 1880,	3	233
do		do	2d do	Dec., 1880,		
do		Philadelphia, Pa.,		March, 1885,	1	208
do		Weston Factory, Newark, N. J.,	1st order,	Oct., 1880,	5	389
do		do	4th do	Sept., 1887,		
BRUSH ELECTRIC LIGHT CO., Philadelphia, Pa.,				July, 1881,	4	300
BRUSH-SWAN ELECTRIC LIGHT CO., Auburn, N. Y.,				Sept., 1885,	1	60
ELECTRIC CLUB, New York, N. Y.,				June, 1887,	1	74
HARLEM LIGHTING CO., New York, N. Y.,				Sept., 1887,	1	300
WESTINGHOUSE ILLUMINATING CO., Schenectady, N. Y.,				Oct., 1887,	2	292
ALLEGHENY COUNTY ELECTRIC LIGHT CO., Pittsburg, Pa.,				Jan., 1888,	2	365
HOUSE OF REPRESENTATIVES, Washington, D. C.,				1887,	1	82
BUCYRUS ELECTRIC LIGHT CO., Bucyrus, Ohio,				June, 1887,	1	85
ST. JOSEPH ELECTRIC LIGHT CO., St. Joseph, Mo.,			1st order,	July, 1883,	2	102
do		do	2d do	Aug., 1884,		
SIR COUTS LINDSAY & CO., Grosvenor Gallery, London, Eng.,				Oct., 1886,	4	956
CARDOGAN ELECTRIC LIGHT CO., London, Eng.,				Oct., 1887,	2	208
THE SCHMIDT-DOUGLASS ELECTRIC LIGHT CO., Limited, Huskegate, Bradford, Eng.,				2 orders, 1887,	2	235
RESIDENCE OF MR. BRYANT, Dorking, England,				Sept., 1885,	2	25
RESIDENCE OF LORD ROTHSCHILD, Tring Park, Herts, Eng.,				June, 1887,	1	60
ELECTRICITEITS MAATSCHAPPY, SYSTEM DE KHOTINSKY, Rotterdam, Holland,				Oct., 1884,	2	164
COMPAGNIE FRANCAISE L'ECLAIRAGE ELECTRIQUE, Paris, France,				Sept., 1887,	2	382
E. LAMY, P. RIEU & CO., Mende, France,				June, 1887,	2	122
A. GILLIBERT & CO., Marseilles, France,				Oct., 1887,	2	220
SOCIETE PER l'ILLUMINAZIONIE ELETTRICA, Palermo, Italy,				Sept., 1887,	2	164
VIENNA OPERA HOUSE, VIENNA, Austria,				Nov., 1887,	6	744
FRANCISCO DE LA VIESCA, Cadiz, Spain,			1st order,	Sept., 1886,	2	102
do		do	2d do	April, 1887,		
CAMELA G. LAGANA, Palermo, Sicily,			1st order,	Sept., 1886,	2	122
do		do	2d do	Aug., 1887,		

NEW YORK, 30 Cortlandt St.   GLASGOW, SCOTLAND, 107 Hope St.
Branch Offices:

BOSTON, Mass., - - - - - - 8 Oliver Street	LONDON, England, - - - 114 Newgate Street
PHILADELPHIA, Pa., - - 32 North 5th Street	MANCHESTER, England, - 3 Victoria Building
CHICAGO, Ill., - - - - 45 South Jefferson Street	PARIS, France, - - - - 20 Boulevard Voltaire
NEW ORLEANS, La., - - 57 Carondelet Street	HAVANA, Cuba, W. I., 116½ Calle de la Habana

THE BABCOCK & WILCOX BOILER.

The Result of 20 Years' Experience on 350,000 Horse Power.

MECHANICAL CONSTRUCTION.

The Babcock & Wilcox Boiler consists of a plain cylinder boiler, serving as a steam and water reservoir, placed above and connected at each end with a nest of inclined heating tubes also filled with water. The rear and lower end of these tubes is connected to a mud-drum at the point furthest removed from the fire. The heat is applied to one-half of the cylinder and all the tube surface.

Dry steam is made, and, therefore, no superheating surface is necessary.

Every square inch of the boiler, inside and out, is in sight, and accessible for mechanical cleaning through a manhole in the cylinder, handholes in the mud-drum, and handholes having milled faces opposite each end of every tube for the interior, and through cleaning doors in the walls for the exterior surfaces.

All joints between the several parts are made by expanding tubes into taper seats, and increased pressure tends to increased tightness.

OPERATION.

The boiler setting forms a furnace in which all the heating surfaces are enveloped by the hot products of combustion as they rise from the grates situated under the front and highest end of the tubes, passing at right angles across them three times and once under the whole length of the cylinder, before being discharged into the stack, at a greatly reduced temperature.

The greater portion of the heat is transferred to the water during the first passage of the gases across the tubes and while combustion is being completed in the triangular chamber under the cylinder.—these being properly fire-box surfaces. The remaining heat is taken up during the second and third passes across the tubes, which act as economizers.

As the water inside the tubes becomes heated, a mingled stream of steam and water is discharged into the front end of the cylinder above, where the steam gradually separates from the water, the latter flowing to the rear end of the cylinder and down again into the tubes, making a rapid and continuous circulation of all the water in the boiler, keeping all parts at a uniform temperature, and avoiding strains from unequal expansion. This rapid circulation also serves to sweep away the steam bubbles from the heating surfaces as fast as formed, supplying their place with water, thus increasing the efficiency of the surface, it also serves to carry any sediment contained in the water into the mud-drum at the rear and lowest point in the boiler, from whence it can be blown out.

The steam is taken out at the top of the steam drum at the rear end.

THE BABCOCK & WILCOX WATER-TUBE BOILER

has all the elements of safety, in connection with its other characteristics of economy, durability, accessibility, etc. Being composed of wrought iron tubes, and a drum of comparatively small diameter, it has a great excess of strength over any pressure which it is desirable to use. As the rapid circulation of the water insures equal temperature in all parts, the strains due to unequal expansion cannot occur to deteriorate its strength. The construction of the boiler, moreover, is such, that should unequal expansion occur under extraordinary circumstances, no objectional strain can be caused thereby, ample elasticity being provided for that purpose in the method of construction.

In this boiler, so powerful is the circulation, that as long as there is sufficient water to about half fill the tubes, a rapid current flows through the whole boiler; but if the tubes should finally get almost empty, the circulation then ceases and the boiler might burn and give out; by that time, however, it is so nearly empty as to be incapable of harm if ruptured.

Charles Munson Belting Co.

MANUFACTURERS OF

EAGLE AND DYNAMO BELTING.

We gladly testify to the superiority of the MUNSON EAGLE and DYNAMO BELT. The stretch, otherwise than the elasticity is removed, no rivets and perfect evenness in heft. The result, noiseless and with a steady motion, which is absolutely essential for a steady light. We use no other make. The character and responsibility of the house is unquestioned.

CHICAGO EDISON LIGHT CO...Chicago.
THOMSON-HOUSTON ELECTRIC LIGHT CO..........................Chicago.
        S. A. BARTON, Gen'l Manager.
BRUSH ELECTRIC CO ....Chicago.
        ALEX. KEMP, Special Agent.
EXCELSIOR ELECTRIC CO...Chicago.
        F. W. HORNE, Western Manager.
MATHER ELECTRIC LIGHT CO...Chicago.
ST. JOE ELECTRIC LIGHT AND POWER CO......................St. Joe, Mo.
        J. A. CORBY.
NATIONAL ELECTRIC CONSTRUCTION CO..........42 La Salle Street, Chicago.
EDISON LIGHT CO..New Orleans, La.
        WM. OSWALD, Agent.

In addition to the above, we are pleased to refer to the following more recent customers:

ALLEGHENY COUNTY LIGHT CO..Pittsburgh, Pa.
THOMSON-HOUSTON ELECTRIC LIGHT CO........................Omaha, Neb.
ST. PAUL GAS AND ELECTRIC LIGHT CO...........................St. Paul, Minn.
CHICAGO ARC LIGHT AND POWER CO...............................Chicago.
LEONARD & IZARD CONSTRUCTION CO............................Chicago.
CALIFORNIA ELECTRIC LIGHT CO....................................San Francisco, Cal.
LOUISIANA ELECTRIC LIGHT AND POWER CO................New Orleans, La.

CHARLES MUNSON BELTING CO.,
28, 30, 32, 34, 36 South Canal Street, CHICAGO.

SAN FRANCISCO,   PITTSBURGH,
29 & 31 Spear Street.   204 Smithfield Street.

NON-MAGNETIC
Watches.

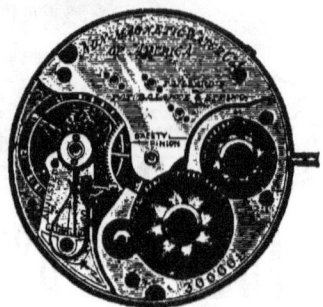

(PAILLARD'S PATENTS.)

THESE Watches contain Paillard's Patent Non-Magnetic Compensation Balance and Hair Spring, and have Non-Magnetic Escapements.

They will not stop or be in any way affected by Magnetism, even when placed in *actual contact* with dynamos or powerful electro magnets.

Every Watch is adjusted to temperature, and for durability, construction, finish, and time-keeping qualities, are unsurpassed.

Endorsed by Prof. Edison, Elihu Thomson, Prof. Edwin J. Houston, N. S. Possons, and others.

FOR SALE BY ALL LEADING JEWELERS.

The Standard Electrical Periodical.

THE
Electrical Review

Published Every Week at

13 PARK ROW, NEW YORK.

The most complete and Reliable Electrical Journal in the World. It is a publication of character; able, impartial, progressive. Graphically Illustrated. PRICE, $3.00 per year, in advance. In the Completeness and value of its Electric Light News and Information it is Unequalled.

GEO. WORTHINGTON, Editor.   **CHAS. W. PRICE, Associate Editor.**

What our Contemporaries say of the ELECTRICAL REVIEW:

Its editorial utterances have been able, independent and fearless, and it has disseminated, regardless of cost, the investigations and experiments of the best minds, and faithfully and intelligibly chronicled each step of progress made in electrical research.—*Modern Light and Heat.*

Its typographical appearance and the excellence of its illustrations are far in advance of any publication issued, either in this country or in Europe—not excepting the Illustrated London News, nor Graphic.—*Boston Commercial.*

The ELECTRICAL REVIEW inaugurates its thirteenth volume with a new cover from which advertising matter is banished. In this it sets an example to the more distinctively literary publications. It is needless to add that its interior evinces corresponding evidences of taste.
— N. Y. Herald.

The REVIEW is a well edited and able paper.
—N. Y. Tribune.

One of the most valuable and most instructive and interesting journals of the day is the ELECTRICAL REVIEW, published weekly in this city.—*N. Y. Sun.*

It appreciates and practices progress.
—*American Machinist.*

Too much praise cannot be awarded to the management of the New York ELECTRICAL REVIEW for the lavish manner in which their Journal is got up for the attraction of students of our fascinating science.
—*London Telegraphist.*

Electrical Review Patent Bureau.

We conduct a general business in the preparation and prosecution of Applications for Patents, in the United States and all Foreign Countries.

Electrical Patents a Special Feature.

All business receives the direct personal supervision of the Manager, whose fourteen years' experience as Patent Attorney, and seven years as Practical Electrician, insures thorough and satisfactory work.
Communications are strictly confidential.

T. J. McTIGHE, Mangr.

Send for our New Descriptive Catalogue of Electrical Books.

Cleveland's Electric Light Cut-Outs

GANG SWITCHES,
From 5 to 40 Amperes.
Quick Make and Break, Uncontrolled by Handle.

CORRESPONDENCE FROM ELECTRIC LIGHT COMPANIES SOLICITED.

FOREST CITY ELECTRIC WORKS,
W. B. CLEVELAND, Electrical Engineer,
183 Seneca Street, - - - CLEVELAND, OHIO.

THE

STANDARD CARBON COMPANY,

MANUFACTURERS OF

Electric Light Carbons

FOR EVERY SYSTEM OF

ELECTRIC LIGHTING,

—ALSO—

Battery Materials.

CLEVELAND, OHIO, U.S.A.

TATHAM & BROTHERS

——MANUFACTURERS OF——

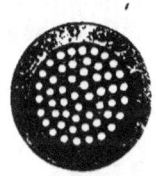 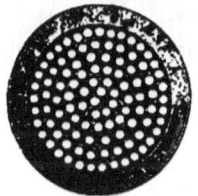 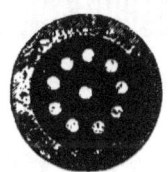

52-Wire Anti-Induction Telephone. 104-Wire Anti-Induction Telephone. 10-Wire Telegraph.

LEAD-ENCASED ELECTRIC

WIRES

◀·············· AND ··············▶

CABLES

2-Wire.

 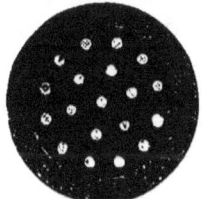 

12-12 Stranded Electric Light. 19-Wire Telegraph. 7-12 Stranded Electric Light.

226 & 228 So. Fifth Street,

PHILADELPHIA.

Special attention given to the manufacture of UNDERGROUND ELECTRIC LIGHT CONDUCTORS, insulated with TATHAM'S PATENT COMPOUND, combining the HIGHEST INSULATION with FLEXIBILITY, SMOOTH FINISH, UNIFORMITY, CONTINUITY and SUPERIORITY OF CONSTRUCTION.

JESSE M. SMITH,

Mechanical and Electrical Engineer,

36 MOFFAT BLOCK,

DETROIT, - - MICH.

Consultation on all Mechanical and Electrical Questions.
Plans and Specifications for Power and Electrical Plants.
Tests of Engines and Electrical Apparatus for Economy and Efficiency.
Inspection of and Reports on Existing Plants.
Expert Witness in Patent Suits.

THE STELLAR ELECTRIC CO.
MANUFACTURERS OF
THE STELLAR ELECTRIC LAMP.

This Company manufactures INCANDESCENT ELECTRIC LAMPS solely. It is the only company in the United States whose business is confined to this industry alone. It has no affiliation with any special electric lighting system, but is prepared to furnish lamps, of superior quality, for any or all systems.

Its special claims are: Unsurpassed Brilliancy, Great Strength of Filament and Long Life, Undiminished Candle Power while In Use, and the Highest Efficiency.

S. K. BAYLEY, General Manager,
Offices, 13 Doane Street, BOSTON, MASS.

THE VULCAN WIRE
FOR THE
ELECTRIC LIGHT, TELEPHONE AND TELEGRAPH.
Manufactured and Sold by

The Campbell Electrical Supply Co.,
95 MILK ST., BOSTON, MASS.

[LIGHT, HEAT AND POWER! May its Light continue to illume the minds, its Heat to warm the hearts, and its Power to draw into closer fellowship the lighting fraternity. Let its Light be as sunshine to the growing grain of inquiry, its Heat consume the chaff of charlatanry, and its Power garner the ripe fruit of honest investigation. EMERSON McMILLAN.]

LIGHT, HEAT AND POWER

THE INDEPENDENT GAS JOURNAL OF AMERICA.

PUBLISHED SEMI-MONTHLY AT PHILADELPHIA.

LIGHT, HEAT AND POWER

Is the only Independent Journal in this Country especially devoted to the Gas Industries. It is newsy, and aims to give a digest of events in the gas world as they occur, without prejudice.

It is the only Gas Journal giving due attention to the developments of the Natural Gas Interests; and the only one in which can be found or had, free discussion, or honest presentation, of the Great Questions of Fuel Gas, from standpoints without bias.

It is the only technical journal in America that has consistently and persistently advocated the

UNION OF GAS AND ELECTRIC LIGHTING INTERESTS.

It has no Personal Axe to Grind; no interests to advocate save those of the general gas public.

In the matter of circulation, this journal claims only what it has; a larger general circulation than that of any other gas journal published in this country, and the best circulation among the legitimate Gas Industries of America.

SUBSCRIPTION :—United States and Canada, $3.00 per year in advance (postage paid). Foreign subscription (Postal Union), 16 shillings, or 20 francs. Single copies, 15 cents. Remittance should be made by P. O. Money Order, Draft or Registered Letter.

ADVERTISING TERMS :—One rate, made known on application.

LIGHT, HEAT AND POWER,

413 Walnut Street, - - PHILADELPHIA, PA.

Western Electrician

PUBLISHED WEEKLY
—AT—
6 LAKESIDE BUILDING,
CHICAGO:
THE ONLY ELECTRICAL JOURNAL PUBLISHED IN THE WEST.

THE WESTERN ELECTRICIAN is the *handsomest, brightest* and *best* Electrical Journal in the world.

Its *descriptive articles* and *illustrations* cover the *new inventions* of America and Europe.

It is *replete* with the *electrical news* of the day.

Its *engravings* and *illustrations* are the *handsomest* published in any Electrical Journal.

SEND $3.00 AND TRY IT FOR A YEAR.

Municipalities, City Councils and **Corporations** desiring to **advertise for bids for Electrical Apparatus**, will find the WESTERN ELECTRICIAN the *very best medium.*

Address:

Western Electrician,
6 LAKESIDE BUILDING,
CHICAGO.

ELECTRICAL ACCUMULATORS
—) OR (—
STORAGE BATTERIES,
—FOR—

Central Station Lighting, Isolated Lighting,
Railroad Car Lighting, Long Distance Lighting
Street Car Propulsion, &c.

THIS SYSTEM ASSURES

Certainty of Light, Steadiness of Current, Economy in
Running Expenses, Extension of Life of Lamps,
Continuity of the Electric Service,

AND CONSEQUENTLY

THE COMPLETE DISPLACEMENT OF GAS.

For Full Particulars address,

THE ELECTRICAL ACCUMULATOR CO.,
44 Broadway, New York.

J. B. YOUNG, Prest. and Treas.   B. K. JAMISON, Vice-Prest.

SOLAR CARBON AND MANUFACTURING CO.

WELL SELECTED   MADE FROM

Electric   Natural Gas.
Light
Carbons
—AND—   Process
Butting Plates,   Patented.

Office, Room 69 Semidt & Friday Building,   **PITTSBURGH, PA.**

M. R. MUCKLE, JR.   T. CARPENTER SMITH.   JOHN S. MUCKLE.

M. R. MUCKLE, JR. & CO.,
608 Chestnut Street, - PHILADELPHIA, PA.

Consulting and Contracting Mechanical and Electrical Engineers.

Steam and Electric Lighting Installations, Isolated and for Central Stations.

Most complete system of Underground Conduits for Electrical Conductors and House Distribution.

WRITE FOR DESCRIPTIVE CIRCULAR.

GEO. WESTINGHOUSE, JR., Pres't.　　　JOHN CALDWELL, Treasurer.
H. M. BYLLESBY, Vice-Pres't.　　　　 A. T. ROWAND, Secretary.
　　and General Manager.　　　　　　 W. L. McCULLAGH, Auditor.

⇁✦THE✦↼

WESTINGHOUSE

Electric Company

PITTSBURGH, PA., U. S. A.

The IDE ENGINE

THE MOST SIMPLE, DURABLE AND ECONOMICAL
AUTOMATIC CUT-OFF ENGINE
IN THE WORLD!

References furnished from the most successful Electric Light Plants in the United States.

Medal and Highest Award from Franklin Institute, of Philadelphia, Pa.

Weitmyer Patent Furnace
FOR BURNING CHEAP FUEL, SCREENINGS, &C.

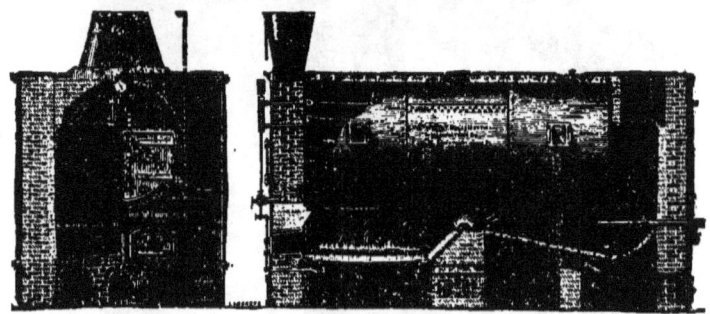

MANUFACTURERS of BOILERS OF ALL KINDS and CONTRACTORS FOR COMPLETE STEAM PLANTS.

MANUFACTURED BY

FOUNDRY and MACHINE DEPARTM'T
Harrisburgh Car Mfg. Company,

HARRISBURGH, - - PA.

L. P. Rider, President.   W. T. Wallace, Treasurer.   W. E. Patrick, Secretary.

Office, No. 52 Fifth Avenue,

Capital Stock, $500,000.   **PITTSBURGH, PA.**

Rider Garbage Furnace Co.
(LIMITED.)

MANUFACTURERS OF THE

Rider Garbage Furnace

FOR CREMATING THE REFUSE OF CITIES.

It can be profitably used in connection with Electric Light Plants.

BOARD OF HEALTH ENDORSEMENT.

Pittsburgh, Pa., January 31, 1888.

To Whom it may Concern:

Having contracted with the Rider Garbage Furnace Company of this city, to furnish and build one of their thirty (30) ton Garbage Furnaces for the use of our city, which, being completed and having given it some three months' trial, we have accepted and paid for it. We are prepared to say about this furnace: It has far exceeded the contract, having destroyed fifty-four (54) tons of garbage in 24 hours, thus surpassing our most sanguine expectations. We add, further, that no other furnace has ever made, so far as we know, such a record here or elsewhere, as this one has in the destruction of garbage, and therefore we strongly recommend its use by other cities as a good sanitary measure.

Board of Health:
- Dr. J. C. Dunn, President.
- Dr. J. D. Thomas, Secretary.
- Crosby Gray, Health Officer.

- W. C. Reitz.
- H. P. McCullough.
- Julius Vortter.
- Dr. James McCann.

WRITE FOR PARTICULARS.

BRADY
Improved Mast Arm

The Simplicity of Construction, Durability, Cheapness and Ease and Quickness of Working, has given

The Brady Improved Mast Arm

The Premium, as the Best Arm in the Market.

HOODS & STORM PROTECTORS

AT LOWEST RATES.

Address,

T. H. BRADY,

NEW BRITAIN, CONN., U. S. A.

Detroit Electric Tower Co.

DETROIT, MICHIGAN.

Descriptive Catalogue and Prices Furnished on Application.

We make all Styles, both Straight and Tapered, but recommend for all purposes our

STANDARD TOWER,

Specially devised for CITY USE, after large experience.

BECAUSE—

1. IT OCCUPIES ONLY THE SPACE OF A LAMP POST.
2. CHILDREN CANNOT CLIMB UPON IT.
3. IT HAS AN ELEVATOR FOR LAMP TRIMMER.

BRUSH CO., DETROIT, 1888: "It is just the thing for cities and towns. Unlike the spread tower it occupies only the space of a lamp post at a street corner. We are greatly annoyed by boys climbing upon our one tapered tower, but they cannot climb upon the Standard Tower.

4. LIGHTS ALLEYS, BACK YARDS AND ALL SPACES.
5. IS PROOF AGAINST STORMS AND EARTHQUAKES.

BRUSH CO., DETROIT, 1888: "We have used 122 Standard Towers for four years past, and although the city has been several times subjected to storms, the most severe in its history, not a cent's worth of damage by the elements has ensued to any of them. Have never had to tighten a joint or apply a tool to any of them."

EVANSVILLE, IND., AFTER CYCLONE 1884: "Houses blown down, trees by the hundred taken out by the roots, not even a rod bent on any of the towers, they defy all elements combined."

SAVANNAH, GA., AFTER EARTHQUAKE 1886: "I found on a close examination that the paint had not started on any of the couplings; that there was no indication of any strain whatever, and the same in regard to all parts of the towers."

6. MOST EFFICIENT SYSTEM, BETTER AND CHEAPER THAN ELECTRIC POLE OR INTERSECTION LIGHTS, GAS OR NAPHTHA.

MAYOR OF DETROIT, 1887: "The tower system I think comes nearer to fulfilling the requirements of efficiency and economy than any mode of street lighting yet introduced. Pole lights at intersections are not a success. We now have 122 towers with four lights each, and I believe they illuminate nearly, if not quite, as much space as four thousand single lights would."

COUNCIL BLUFFS, IOWA, 1887: "Fine, as good as moonlight." "An unqualified success." "A grand success."

The Towers of this Company are now used in about Thirty Cities.

Address,

DETROIT ELECTRIC TOWER CO.,
DETROIT, MICH.

G. W. Stockly, - - - President.
J J Tracy, Vice-Prest.   J. Potter, Treas.   N. S. Possons, Supt.
Wm. F. Swift, Sec'y.   S. M. Hamill, Jr., Ass't Sec'y.   W. J. Possons, Ass't Supt.

THE

BRUSH ELECTRIC COMPANY

OF CLEVELAND, OHIO.

MANUFACTURERS OF

BRUSH ARC

—AND—

INCANDESCENCE

ELECTRIC

Lighting Apparatus.

ELECTRIC MOTORS,

Carbons for Arc Lamps, Etc., Etc.

THE
Globe Carbon Co.
CLEVELAND, O., U. S. A.

CARBONS
FOR
ELECTRIC
LIGHTING,
BATTERIES,
MOTORS,
ETC.

WE BEAT
THE WORLD
FOR
BRILLIANCY,
STEADINESS,
AND
LONGEVITY.

TRY OUR CARBONS!
IF THEY DO NOT PROVE SATISFACTORY, YOU MAY HOLD SAME SUBJECT TO OUR ORDER.

☞ **Write Us for Prices.**

Chas. A. Cheever,
Prest.

Willard L. Candee,
Treas.

TRADE MARK.

The Okonite Company
13 PARK ROW, NEW YORK,
MANUFACTURERS OF

Electric Light, Telephone and Telegraph Wires and Cables
For AERIAL, SUBMARINE and UNDERGROUND USES.

Sole Manufacturers of the Celebrated **OKONITE TAPE**

The only Absolute and Safe Waterproof and Insulating Tape in the Market.

AGENTS AT Boston, Philadelphia, Chicago, Minneapolis, Kansas City, Cincinnati and Louisville.

NEW YORK
Electric Construction Co.

242 & 244 East 122d St.,

NEW YORK CITY.

ANDREW L. SOULARD, President.
JOHN H. HAPGOOD, Vice-President.
E. F. AMES, Secretary.

F. G. CARTWRIGHT, Supt. of Construction.
GEO. H. REYNOLDS, Consulting Engineer.
M. M. SLATTERY, Consulting Electrician.

CONTRACTS TAKEN

—FOR THE—

INSTALLATION OF ALL ELECTRICAL PLANTS

ESTIMATES FURNISHED

For Electric Light and Steam Plants Complete.

SEND FOR CIRCULAR.

Estimates Promptly Furnished for Erecting Electric Lighting
Plants for Cities, Companies or Individuals.

JENNEY ELECTRIC COMPANY,

DANIEL W. MARMON, President.
ADDISON H. NORDYKE, Vice President.
BRAINARD ROBBOIS, Secretary.
AMOS K. HOLLOWELL, Treasurer.
CHARLES D. JENNEY, Electrician.

OFFICE AND WORKS:
Cor. Kentucky Ave. and Morris St.,
INDIANAPOLIS, INDIANA.

Improved Dynamo Lamp & Electric Motor.

Sole Owners of all the Patents and Inventions of Charles D. Jenney (known as the Jenney System) and Sole Owners and Manufacturers of the

In all desirable features of Arc and Incandescent Lighting the Jenney System leads. Simple, durable, economical, steady, brilliant and penetrating. In these essentials it challenges comparison.

The Jenney Incandescent dynamos are self-regulating, and permit the turning on and off of one or all of the Lamps at will.

☞ Send for Pamphlets Illustrating and describing the system.

Prices Furnished for the JENNEY ARC or INCANDESCENT SYSTEMS, or for both Combined.

WESTERN ELECTRIC ✢ COMPANY

Chicago.   New York.
London.   Antwerp.

Electric Lighting Apparatus

OUR SYSTEM OF ELECTRIC LIGHTING IS COMPLETE IN EVERY DETAIL, AND IS UNEXCELLED IN THE EFFICIENCY AND WORKMANSHIP OF THE APPARATUS, IN THE BRILLIANCY, STEADINESS AND COLOR OF THE LIGHT PRODUCED, AND IN AUTOMATIC REGULATION. WE FURNISH DYNAMOS TO GIVE ANY REQUIRED CURRENT.

The high grade of workmanship and the efficiency of our Dynamos render them especially advantageous for charging Storage Batteries, furnishing current for Electric Motors, etc. We furnish Arc Lamps adapted to any Arc Current. The superior construction and simplicity of our Lamp render its use advantageous with any system.

We manufacture and deal in all kinds of Electrical Apparatus and Supplies for **FIRE** and **POLICE** service.

www.ingramcontent.com/pod-product-compliance
Lightning Source LLC
Chambersburg PA
CBHW032148230426
43672CB00011B/2492